T0273624

The Mathematics of Ciphers
Number Theory and RSA Cryptography

S. C. Coutinho

Department of Computer Science
Federal University of Rio de Janeiro
Rio de Janeiro, Brazil

CRC Press
Taylor & Francis Group
Boca Raton   London   New York

CRC Press is an imprint of the
Taylor & Francis Group, an **informa** business

AN AK PETERS BOOK

First published 1999 by A K Peters, Ltd.

Published 2019 by CRC Press
Taylor & Francis Group
6000 Broken Sound Parkway NW, Suite 300
Boca Raton, FL 33487-2742

© 1999 by Taylor & Francis Group, LLC
CRC Press is an imprint of Taylor & Francis Group, an Informa business

First issued in paperback 2019

No claim to original U.S. Government works

ISBN 13: 978-0-367-44760-1 (pbk)
ISBN 13: 978-1-56881-082-9 (hbk)

Visit the Taylor & Francis Web site at
http://www.taylorandfrancis.com

and the CRC Press Web site at
http://www.crcpress.com

Library of Congress Cataloging-in-Publication Data

Coutinho, S. C.
   [Números inteiros e criptografia RSA. English]
   The mathematics of ciphers : number theory and RSA cryptography
/ S.C. Coutinho.
   p. cm.
Includes bibliographical references and index.
ISBN 1-56881-082-2
1. Number theory. 2. Cryptography. I. Title
QA241 .c69513 1998
512′.7--dc21

98-49611
CIP

For Andrea and Daniel

And I wish the Reader *also to take notice, that in writing of it I have made myself a recreation of a* recreation; *and that it might prove so to him, and not read* dull *and* tediously, *I have in several places mixt (not any scurrility, but) some innocent, harmless mirth; of which, if thou be a severe, sowre-complexion'd man, then I here disallow thee to be a competent judge; for* Divines *say,* There are offences given, and offences not given but taken.

Izaak Wolton in "The Compleat Angler"

Contents

Preface

This book will take you on a journey whose final destination is the celebrated Rivest, Shamir, and Adleman (RSA) public key cryptosystem. But it will be a leisurely journey, with many stops to appreciate the scenery and contemplate sites of historical interest.

In fact, the book is more concerned with mathematics than with cryptography. Although the working of the RSA cryptosystem is described in detail, we will not be concerned with details of its implementation. Instead, we concentrate on the mathematical problems it poses, which are related to the factorization of integers, and to determining whether a given integer is prime or composite. These are, in fact, among the oldest questions in the area of mathematics known as *number theory*, which has been a wonderful source of very challenging problems since antiquity. Among the mathematicians that worked in the theory of numbers we find Euclid, Fermat, Euler, Lagrange, Legendre, Gauss, Riemann, and, in more recent times, Weil, Deligne, and Wiles.

The way number theory is presented in this book differs in some important respects from the classical treatment of most older books. Thus we emphasize the algorithmic aspects everywhere, not forgetting to give complete mathematical proofs of all the algorithms that appear in the book. Of course number theory has been permeated by algorithms since the time of Euclid, but these had been, until very recently, somewhat out of fashion. We take the algorithmic approach very seriously. Hence, Euclid's proof of the infinity of primes is preceded by a discussion of the primorial formula for primes, and the existence of primitive roots modulo a prime number is proved using the algorithm Gauss invented for computing the roots.

Hence, this is really a book about algorithmic number theory and its applications to RSA cryptography. But, although the book has a very sharp focus, the presentation is by no means narrow. Indeed, we have not always followed the most direct path, but chosen instead the one that promised to throw most light onto the subject. Hence the introduction of groups, which allow us to give a unified treatment of several factorization methods and primality tests in Chapters 9 and 10. Our excursion into group theory will take us as far as Lagrange's theorem, and includes a discussion of groups of symmetry.

This book grew out of lecture notes aimed at first-year students of computer science. Some of the features of the book are really a consequence of the poor background knowledge of the students. Thus very little previous knowledge of mathematics is required. Indeed, we never use anything beyond geometrical progressions and the binomial theorem. Also, although the book is mainly concerned with algorithms, no knowledge of programming is assumed. However, given the subject matter of the book, it is to be expected that many readers will be computer literate. Thus every chapter ends with some (optional) programming exercises that illustrate the algorithms presented in the text. Many of these are aimed at generating numerical data to be compared with some known conjecture or formula of number theory. Thus they are of the sort that is sometimes called a *mathematical experiment*.

A comment on style: Books on mathematics are sometimes written as a dry sequence of definitions, theorems, and proofs. This style goes back to Euclid's *Elements*, and it became the standard way to present mathematics in the late twentieth century. But we should not forget that this marmoreal style did not prevail even among the Greek mathematicians. Thus Archimedes tells his readers of his difficulties and wrong turns, and he even warns his readers against propositions that he used and then discovered to be false. In this book I prefer to follow Archimedes' example rather than Euclid's. This choice affected the style of the exposition in many respects. First, the historical comments are part of the text, and not isolated notes, and range from the origins of group theory to mere tales of eccentricity. Second, the algorithms are described as instructions in plain English, and I made no effort to optimize them, unless that would promote understanding. Five years of teaching this material have convinced me that this is no obstacle to programming the algorithms for those with the necessary background.

Another characteristic of the book that should be mentioned is that the most important theorems and algorithms are referred to by name. Most of these names are classical, and have been used to identify the results for decades, even centuries. Where a name wasn't already available I made one up. Some, like the *primitive root theorem*, will be immediately identified by those who are familiar with the subject; some may be harder to identify. Since this makes dipping in more difficult, I have provided a separate index of all the *main* theorems and algorithms, which also includes a short description of their content. The results that do not have special names are referred to by the chapter and section where they can be found.

This is a revised version of a book first published by IMPA and the Brazilian Mathematical Society in Portuguese in 1997, and it grew out of my lecture notes for the first-year students of computer science at the Federal University of Rio de Janeiro. I owe more to the students who took the course during these last five years than I can possibly thank them for. They influenced both the style of the presentation and its content, and their suggestions and criticism helped me in correcting mistakes and simplifying the presentation of many proofs.

I am especially grateful to Jonas de Miranda Gomes. It was he who first thought that the book deserved an English version, and he also made all the necessary contacts. This book wouldn't even have been conceived if it weren't for him. Many thanks also to Amilcar Pacheco, Martin Holland, and Keith Matthews for their suggestions and comments.

Finally, I wish to thank all the people at A K Peters I have been in contact with as the book developed. Their support and patience, even at the times when I felt most discouraged, helped me carry the work forward to its conclusion.

Rio de Janeiro, July 18, 1998

Introduction

The main character of the story this book tells is the public key cryptosystem known as RSA. All the mathematics that we will study is, directly or indirectly, concerned with it. The introduction contains a brief (and somewhat incomplete) description of the RSA, and a short history of the area of mathematics that serves as its foundation: number theory.

1. Cryptography

In Greek, *cryptos* means "secret, hidden". *Cryptography* is the art of disguising a message so that only its legitimate recipient can understand it. There are two sides to this process. The first is the procedure by which the original message, or *plaintext*, is disguised. This is called *encryption*, and the encrypted message is known as a *cryptogram*. The legitimate recipient of the message must know an inverse process by which the cryptogram is translated back into the original message. This is called *decryption*.

Many of us played with simple ciphers when we were children. The most common of these consists of replacing each letter of the alphabet by the one that follows it, and replacing Z by A. This is essentially the kind of cipher that Caesar used to exchange secret messages with the Roman generals throughout Europe.

Cryptography has a twin sister, called *cryptoanalysis*, which is the art of breaking a cipher. Of course, to "break" a cipher is to find a way to decrypt it when one does not hold the decryption key; that's what eavesdroppers have to do. Ciphers like Caesar's are very easy to break. Indeed any encryption procedure that works by the systematic substitution of the letters of the alphabet with other symbols suffers from the same weakness. This is due to the fact that the average frequency of each letter in a given language is more or less the same. For example, the most frequently used letters in English are *E*, *T*, and *A*. Thus it is possible to guess the letters that correspond to the most often used symbols in a cryptogram that was encrypted with a substitution cipher. Moreover, the most frequently used words in English are *the* and *and*. So the groups of three characters that appear most often in the cryptogram will probably correspond to one of these words. And so on. This strategy is known as *frequency analysis*.

Note that these comments assume that the message is reasonably long. One can always write a short message whose letter and word counts are as unlikely as one wishes. However, this is almost impossible with a long message, because the frequency counts are a characteristic of the language itself.

Cryptoanalysis has other uses besides breaking ciphers, and in these, frequency analysis is also very important. One of these applications is to the decipherment of ancient inscriptions. The most famous is probably that of the Egyptian hieroglyphs by J.-F. Champollion in 1822. The key to the decipherment was the Rosetta Stone, a block of black basalt found in 1799, that is now at the British Museum in London. The stone contains the same text written in three different scripts: hieroglyphic, demotic, and Greek.

In Champollion's time it was widely believed that the Egyptian hieroglyphs were a logographic system of writing. In these systems every symbol corresponds to an idea. Of the writing systems still in use, Chinese comes closest to being logographic. The sages of Champollion's time had good reason for believing the logographic theory. After all, the oldest detailed discussion of the the nature of the hieroglyphs assumed that they were a form of pictorial writing. This work was compiled by one Horapollo of Nilopolis in the fourth or fifth century A.D., a time when the knowledge of how to read and write the old script was essentially extinct.

Champollion decided to check Horapollo's assumption by performing a frequency analysis in the texts of the Rosetta Stone. First he counted the words in the Greek text, and found that there were 486 of them. Thus, if each hieroglyph corresponded to an idea (or word), there ought to be about that many symbols in the hieroglyphic text. But Champollion's count revealed 1419 characters, far more than expected. Thus the hieroglyphs did not constitute a logographic system after all.

Champollion's work did not stop at that, and by 1822 he had finally found the key to deciphering of the writing system of the Ancient Egyptians. We now know that the system is quite complex. It is essentially logo-syllabic, so that a symbol stands either for a word or for the syllable with which that word begins. But that's not all. A symbol can also be used as a determinative. For example, after the name of a person, the Egyptians could add a male or female figure, which allowed the scribe to tell whether the name belonged to a man or a woman. For more details on the hieroglyphic system and on the history of the decipherment, see Davies 1987.

Of course frequency analysis can be considerably sped up with the use of a computer. This is why many of the old methods of encryption have now become obsolete. Let's not forget that some of the first computers were built to break the German codes used during the Second World War.

Nowadays, the communication between computers using the Internet is posing new challenges to cryptographers. Since the messages are sent through telephone lines, it is necessary to encrypt them if they contain any sensitive information. That need not be a great government secret; it could be just your credit card number! Imagine that a company does its bank transactions in this way. Two problems immediately come to the fore. First, it is necessary to make sure it will not be possible to read the message if it is intercepted by an eaves-dropper. Second, the bank must have some way of knowing that the message has originated with a legitimate user at the company. In other words, it must be possible to sign an electronic message.

Many of the ciphers that are used in this new environment are of the kind known as *public key cryptosystem*. These were introduced in 1976 by W. Diffie and M. E. Hellman of Stanford and independently by R. C. Merkle of the University of California. In a public key cryptosystem the fact that one knows how to encrypt a message does not mean that it can be easily decrypted. So anyone can be told how a message that will be sent to a certain bank is to be encrypted—that does not put at risk the security of the system. This has obvious advantages for commercial transactions.

At first this may seem an impossible idea. For example, consider the cipher that consists of replacing each letter by the one that follows it. In this case, a cryptogram is decrypted by replacing each of its letters with the one that comes before it in the alphabet. Thus, for this cipher, knowledge of the encryption process immediately gives access to the decryption procedure. Unlike this cipher, the public key cryptosystems have a "trapdoor", that is, an operation that is easy to perform, but difficult to undo. An example of a trapdoor is considered in the next section, where we give a rough description of the most popular of the public key cryptosystems, the RSA.

2. The RSA cryptosystem

The best known and one of the most widely used of the public key cryptosystems is the RSA. It was invented in 1978 by R. L. Rivest, A. Shamir, and L. Adleman at MIT (see Adleman et al. 1978). A detailed description of the RSA will be given only in the last chapter, because it depends on the ideas and techniques that will be developed throughout the book. However, it is convenient to have some understanding of the reason why knowledge of the encryption procedure of the RSA does not give one immediate access to the decryption process. This is an example of the concept of a trapdoor, mentioned at the end of the previous section.

Suppose that we want to implement the RSA cryptosystem for a given user. The basic ingredients are two distinct odd prime numbers that we will call p and q. Let n be the product of these primes; thus $n = pq$. The *public* or *encryption* key of this user of the RSA is the number n (and another number that we need not worry about now). The *secret* or *decryption* key consists essentially in the primes p and q. Thus every user has a personal pair of primes that must be kept secret. However, the product of these primes (the number n) is made public. Thus, if a bank uses RSA, anyone can send it an encrypted message, because its encryption key is known to all.

Why, then, is it difficult to break the RSA? After all, one need only factor n to find p and q. However, if the primes have more than 100 digits each, the time and resources required to factor n are such that the system becomes very hard to break. Thus the trapdoor of the RSA lies in the fact that it is easy to multiply p and q to get n, while factoring n to get p and q can be next to impossible. In other words, although it is conceptually quite easy to describe how one breaks the RSA, it can be almost impossible to do it in practice. But one must not forget that the obstacle is essentially of a technological kind. In other words, it is conceivable that advances in hardware, and better methods for the factorization of integers, could one day render the RSA obsolete. This point was dramatically illustrated when the RSA-129 was broken. This was a test message encrypted by the inventors of RSA in 1977, and it got its name from the fact that the encryption key had 129 digits. It was calculated that with the methods then available, it would take 40 quadrillion years to factor the key and decrypt the message. The combination of advances in hardware, new factorization methods, and the advent of the Internet led to the factorization of this 129-digit number in 1986. This was done after eight months of work, during which the computers of 600 volunteers in 25 countries were used. Each computer worked on a little piece of the problem during idle computer cycles. All the pieces were later put together using a supercomputer. The advances in this area have been such that the larger RSA-130, factored in 1996, took only 15 percent of the computer time required to factor RSA-129.

Summing up the previous discussion, we conclude that

(1) to implement RSA we need two large prime numbers p and q;
(2) to encrypt a message using RSA we use $n = pq$;
(3) to decrypt an RSA cryptogram we must know p and q; and
(4) the security of RSA depends on the fact that it is difficult to factor n, and find p and q, because these are very large numbers.

When we finally come to a complete understanding of RSA in Chapter 11, you'll see that this is not totally accurate, even though it touches on all the main points.

Have you noticed that (1) and (4) above seem to contradict each other? Indeed, the security of the RSA cryptosystem depends on the difficulty of factoring n. Now n is difficult to factor because it is very large. But $n = pq$ is large when p and q are large. However, we need to prove that the latter numbers are prime, and a prime number is one that does not have any factors except 1 and itself. Thus to prove that p is prime we need to apply to p a factorization algorithm, and make sure that no factors are found, except 1 and p. But don't we then run into the same problem we had with respect to n? If a factorization algorithm will take too long to factor n, won't it take too long to prove that p is prime?

That's indeed the case, but it doesn't mean that we have no way of proving that large numbers are prime. That's because we do not prove primality by trying to factor a number. There are methods for checking whether a number is prime or composite without ever attempting to factor the number. For example, we know that $2^{2^{14}} + 1$ is composite, but none of its factors are known. This is not surprising, since this is a number of 4933 digits. Indirect methods for testing primality and compositeness are discussed in Chapters 6 and 10.

The previous discussion makes clear that in order to explain and implement the RSA cryptosystem, one needs a good knowledge of the properties of the integers. Two problems are especially relevant:

- How does one factor a given integer efficiently?
- How does one prove that an integer is prime?

As we have seen, although these problems are closely related, the second is not a consequence of the first. The area of mathematics that is concerned with the properties of the integers is called *number theory*. It is one of the oldest areas of mathematics, and the problems mentioned above are some of the earliest number theoretic problems considered by humankind.

There is a question of a more practical nature that we must address before we discuss number theory in more detail. The public key used by the RSA cryptosystem is a very large number; in practical applications this number has more than 200 digits. How does one compute with such large numbers? Of course one needs a computer; however, most programming languages will not allow one to deal directly with such huge numbers. The easiest way out is to use a computer algebra system.

3. Computer algebra

Computer algebra is concerned with exact computations with numbers, such as very large integers and fractions, and with symbolic computations with functions, such as polynomials, sines, and cosines. By "exact computation" we mean that the system does not use floating point arithmetic, unless told to do so. For example, in computer algebra, fractions are represented in the form numerator over denominator, and calculations with fractions are done just as with pencil and paper.

From our point of view, the most important feature of a computer algebra system is that it allows one to compute with very large integers. For most of this book we will simply assume that such calculations are possible. In fact, it is not possible to go into the detailed implementation of the methods for representing and calculating with large numbers within the limits of an elementary book like this one. On the other hand, one ought not to be completely in the dark about this point, since it is really behind most of what we will be doing. In this section we take the middle ground: We explain how large positive integers are represented, and discuss how the computer can be programmed to add two such numbers.

This is one of the few places in the book where it will be assumed that you are familiar with computer programming. Since the results of this section are not used anywhere else in the book, you may skip the rest of it with a clear conscience.

Most computer languages allow you to write programs that calculate with integers. The snag is that these integers cannot exceed a certain size. In other words, the standard programming languages behave as if the set of integers were finite and, indeed, rather small. Of course this won't do for our purposes. We must be able to compute with very large integers; moreover, we have no way of knowing in advance what the maximum size of these numbers will be. We get around this problem by writing special software for representing and calculating with large integers. This software will be written in one of the standard programming languages. Thus we will assume that we have a computer that can represent and calculate with integers up to a certain upper bound, which depends on the language we are using. These are called *single precision* integers. The software we want to write will use the operations with single precision integers to represent and calculate with integers of indeterminate size, called *multiple precision* integers.

Needless to say, all computers have finite memory, so the size of the integers that they can handle is always limited. Thus, when we talk of integers of "indeterminate size", what we really mean is that the program can handle big enough integers so that for all practical purposes this upper limit can be ignored.

Recall that we are assuming we have a computer that runs a programming language that can handle integers in single precision. Let b be the largest power of 10 supported as a single precision integer by this language. Then b is the *base* of the number system we will use to represent multiple precision integers. If $n > 0$ is an integer, then there exist unique integers $a_0, a_1, \ldots, a_k$ between 0 and $b - 1$, such that

$$n = a_k \cdot b^k + \cdots + a_2 \cdot b^2 + a_1 \cdot b + a_0.$$

The *positional representation of n in base b* is

$$n = (a_k \ldots a_1 a_0)_b.$$

The integers $a_0, \ldots, a_k$ are the digits of the representation of n in base b. Inside the computer, n corresponds to a list, each node of which stores a base b digit of n. Recall that these digits are single precision integers, so the computer already knows how to calculate with them. The number of nodes of the list will depend on the size of n.

Of course, we do not want only to represent large integers in the computer; we want to calculate with them. The addition of integers written in base b can be carried out just as the addition of integers written in the decimal system. The only thing we have to worry about is the *carry*.

Suppose that we wish to add two positive integers

$$c = (c_k, \ldots, c_1, c_0)_b \quad \text{and} \quad d = (d_k, \ldots, d_1, d_0)_b.$$

In other words, we want to find the base b digits $s_0, s_1, \ldots$ of the base b representation of their sum. We first compute $c_0 + d_0$, which is only a sum of two single precision integers. Now, the least significant digit s_0 of $c + d$ will be the remainder of the division of $c_0 + d_0$ by b. If $c_0 + d_0 < b$, there is no carry, and we can proceed to the next digit. Suppose, however, that $c_0 + d_0 > b$. Since both c_0 and d_0 are smaller than b, it follows that

$$c_0 + d_0 < 2b - 2 < b^2.$$

In particular, the quotient of the division of $c_0 + d_0$ by b cannot be greater than 1. Thus the carry, if it is not 0, is always 1. To compute s_1 we proceed in a similar fashion, but we must not forget the carry from the sum of the previous base b digits. Thus s_1 is the remainder of the division by b of $c_1 + d_1 + carry$. Moreover, if $c_1 + d_1 + carry > b$, then there will be a carry of 1 when we come to compute s_2. And so on, until all the base b digits have been added up.

It is very easy to write software that uses lists to implement the positional representation and the addition of integers in base b. The multiplication also closely follows the usual pencil and paper method. The division, however, presents some interesting problems that are briefly discussed in Chapter 1, section 2.

You should also be aware that writing good software to compute with integers is not restricted to programming procedures for doing arithmetic operations. Consider what happens when we multiply two integers. The usual method produces many intermediate numbers that are really not needed after the product has been obtained. Unless we spot and delete them, they'll remain in the computer, occupying precious memory space. If the numbers are large, the growing accumulation of such "rubbish" can quickly draw the system to a halt. Thus we must have a way of automatically finding and deleting unnecessary numbers.

These comments only touch a very important and rapidly growing area in the interface between computer science and mathematics. For a thorough study of the methods used in the computation with large integers, see Knuth 1981. A shorter account, which also includes a more detailed discussion of the representation of integers as lists, can be found in Akritas 1989.

4. The Greeks and the integers

A knowledge of the integers, and of their fundamental operations, was common to all of the ancient civilizations. Their knowledge was of an almost purely heuristic kind, but it was sufficient since they used numbers mostly for counting and keeping accounts. The Greeks, who had a more philosophical bent, began to consider the integers as independent entities with a life of their own, and not as a mere aid to counting. This led them to distinguish between *logistics* and *arithmetic*. The former was "the science which deals with numbered things, not numbers"; the latter aimed at seeing "the nature of numbers with the mind only". The second quote is from Plato's *Republic*, where we also read that

> arithmetic has a very great and elevating effect, compelling the soul to reason about abstract number, and rebelling against the introduction of visible or tangible objects into the argument.

See Plato 1982, vii.525. Ironically, we now use the word *arithmetic* to describe what the Greeks called logistic. But their arithmetic is not dead; it has been transmuted into our theory of numbers.

Among the number theoretic problems that the Greeks studied, we have

- the method for computing the greatest common divisor of two integers;
- the method for finding the positive primes smaller than a given number; and
- the existence of infinitely many prime numbers.

These problems are discussed in detail in the most famous mathematical book the Greeks bequeathed to us, Euclid's *Elements*, written in Alexandria around 300 B.C.

The *Elements* are divided into 13 books. Three of these deal with number theory; the other books are concerned with plane and solid geometry, and the construction and properties of real numbers. The discussion of number theoretic problems begins in Book VII. In it we find the definitions of prime and composite numbers, and the method for computing the greatest common divisor by successive divisions. Book VIII is mainly concerned with geometric progressions. Book IX contains the proof that there are infinitely many primes, which we discuss in Chapter 3, section 5; and a formula for perfect numbers, which can be found in the exercises of Chapter 2.

Many other Greek mathematicians studied problems of a number theoretic nature. The most important of these was Diophantus. In his *Arithmetic*, written about A.D. 250, he considers in detail the problem of solving indeterminate equations with integer coefficients; see Chapter 4, section 6, for more details. After Diophantus, Greek mathematics moved away from number theory, and although there were many good mathematicians among the Arabs, Indians, and in the Europe of the Renaissance, none of them was directly concerned with number theory. Indeed, the subject was more or less dormant until it was rediscovered, directly from the Greek source, in the seventeenth century.

5. Fermat, Euler, and Gauss

The Renaissance was a time when the works of many Greek authors were rediscovered, edited, and published; but this wasn't true of many of the Greek mathematicians. For example, it was only in 1621 that Bachet published the original text of Diophantus' *Arithmetica*, together with a Latin translation. Sometime before 1636, Pierre de Fermat, a councillor in the Toulouse High Court, acquired a copy of this book. Fermat studied mathematics in his free time, and he read and carefully annotated his copy of Bachet's edition. The ideas prompted in Fermat's mind by Diophantus' work mark the rebirth of number theory in Europe.

Fermat was born in 1601, and he wasn't a professional mathematician. Indeed, few people made a living from mathematics at that time. It was also difficult to make one's ideas known to other mathematicians, because there were no specialist journals. The first journal totally dedicated to mathematics appeared only in 1794. However, as Fermat's contemporary Pascal said, perhaps with some exaggeration, mathematicians "are so few in number as to be unique among a whole people and over a long period of time". Thus it was possible for the mathematicians of the time to communicate their ideas more or less effectively through their correspondence. This was made easier by the fact that some people played the role of intermediaries; as soon as they were told the latest news, they passed it on to their correspondents. In Fermat's time the most famous of these intermediaries was Father Marin Mersenne. His circle

P. de Fermat (1601–1665).

of correspondents included, besides Fermat himself, such luminaries as Pascal, Descartes, and Roberval.

It was in the form of letters to Mersenne and other mathematicians of the time that most of Fermat's work reached us. After Fermat's death, his son Samuel collected whatever he could find of his father's papers with a view to publishing them. The first published volume was Bachet's edition of Diophantus, including all the margin annotations by Fermat. The most famous of all these annotations concerns the statement that later became known as *Fermat's Last Theorem*. In modern parlance it says that if x, y, z, and n are integers such that $x^n + y^n = z^n$ and $n \geq 3$, then $xyz = 0$. In his note, Fermat claims to have a "marvellous proof" of this result but adds that "the margin is not large enough to write it in". Fermat's Last Theorem was finally proved in 1995—more than 300 years after it was first stated. For more details, and references, see Chapter 2, section 8.

Fermat's work in mathematics is not limited to number theory. He also made key contributions to analytic geometry and the integral and differential calculus. Moreover, with Pascal, he is codiscoverer of the calculus of probability. Although Fermat was in love with number theory, he had no success in his attempts to interest his contemporaries. His heir in this respect was Leonard Euler, who was born in 1707, 42 years after Fermat's death.

Euler was one of the most prolific mathematicians of all time, and he contributed to most of the areas of pure and applied mathematics that existed in the eighteenth century. Unlike Fermat, he was paid to work as a research mathematician, and he held posts in the academies of Berlin and St. Petersburg. These academies were in fact research institutions, whose memoirs published the contributions of their members.

L. Euler (1707–1783).

Euler's interest in number theory was a result of his correspondence with Christian Goldbach. Like Mersenne before him, Goldbach wasn't a great mathematician; mathematics was his hobby. However, it was through him that Euler came to Fermat's work in number theory. In his first letter to Euler, dated 1729, Goldbach adds the following postscript:

> Is Fermat's observation known to you, that all numbers $2^{2^n} + 1$
> are prime? He said he could not prove it; nor has anyone else
> done so to my knowledge.

Euler's immediate reaction was one of skepticism, and he doesn't seem to show much interest; but Goldbach insists, and, in 1730, Euler begins to read Fermat's work. In the coming years he would prove and extend a good part of the results stated by Fermat. In Chapter 9 we will study the method by which Euler settled (in the negative!) the question raised by Goldbach. More details of the history of the numbers of the form $2^{2^n} + 1$ can be found in Chapter 3, section 3.

After Euler's work, number theory became far more popular than it had been hitherto. But its systematic development began only with the *Disquisitiones arithmeticæ* of C. F. Gauss, published in 1801. The influence of Gauss's book was enormous, and is witnessed by the fact that most books on the subject, including this one, still follow his approach. Many of the results and techniques that we will study come directly from the *Disquisitiones*.

Gauss was the son of a manual worker, but he was a child prodigy, and his mathematical ability was noticed very early. His contributions to mathematics extend far and wide, and include such diverse fields as differential geometry and celestial mechanics. He also made important contributions to geodesy and physics. Such was the importance of his work that he earned from his contemporaries the title "prince of mathematicians".

Fermat, Euler, and Gauss are the heroes of this book. Most of the ideas we will study originated either in Ancient Greece or in the work of one of these three mathematicians.

6. The problems of number theory

We cannot finish our introduction to number theory without at least trying to explain what makes it so captivating that Gauss declared it the "queen of mathematics". The best way to give an idea of the attractions of this area is to describe some of its problems; these often combine a very simple statement with a proof that is both ingenious and of great technical virtuosity.

We list a few number theoretic questions below. After reading them you might try to guess which are the most difficult to solve, and which the easiest. You'll probably be in for some surprises when you read the present status of each question afterward.

(1) If p is a prime number, does p always divide $2^{p-1} - 1$?
(2) Is there a prime p such that p^3 divides $2^{p-1} - 1$?
(3) Is each even integer greater than 2 a sum of two primes?
(4) Are there two consecutive integers, apart from 8 and 9, which are powers of integers?
(5) Can every odd prime number be written as a sum of two squares of integers?
(6) Are there infinitely many pairs of prime numbers of the form $p, p + 2$?
(7) Are there infinitely many prime numbers p for which $2^p - 1$ is also a prime?

Questions (1) and (2) are very similar. Comparing them, one may even come to the conclusion that (2) should be easier than (1). After all, to settle (2) one need only find *one prime* satisfying a certain property; but to prove (1) one must show that a very similar property holds for *every prime*. The truth, however, is that the answer to (1) has been known to be yes since the seventeenth century, when it was proved by Fermat, while (2) is still an open question.

The third question is a famous conjecture of Goldbach, and though it has been around for more than 200 years, no one has yet managed to prove it. Question (4) is known as *Catalan's conjecture* and is also an open question. However, it is known that if the difference between a cube and a square is ±1, then the cube is 8 and the square is 9. This was shown by Euler, and the proof is not very difficult. For more details on Catalan's conjecture see Ribenboim 1994.

The answer to question (5) is no. If an odd prime p leaves remainder 3 when it is divided by 4, then it is not possible to find integers x and y such that $x^2 + y^2 = p$. This was shown by Fermat, and the proof can be found in exercise 13 of Chapter 4. However, Fermat also showed that if the prime leaves remainder 1 when divided by 4, then the question has an affirmative answer.

More details can be found in exercise 14 of Chapter 5. The history of this problem is discussed in Weil 1987, Chapter II, section VIII.

Problem (6) is the famous *twin-primes conjecture*, and it is not known whether it is true or false. Of course the fact that there are infinitely many primes was known to Euclid, and his proof of it can be found in Chapter 3, section 5. It is also known that if a and r are integers whose greatest common divisor is 1, then there are infinitely many primes of the form $a + kr$, where k is a positive integer. This was proved by L. Dirichlet in 1837. A very particular case of this result is proved in exercises 3 to 7 of Chapter 3. Another related conjecture asks whether there are infinitely many primes p for which $p + 2$ and $p + 6$ are also prime; in this case too the answer is not known. However, there is only one prime p for which $p + 2$ and $p + 4$ are also prime; see exercise 9 of Chapter 3.

The last question is also an open problem. The numbers of the form $2^n - 1$ are called *Mersenne numbers* for a reason that is explained in Chapter 3, section 2. If the exponent n is composite, then so is the number $2^n - 1$. However, if the exponent is prime, the corresponding Mersenne number can be either prime or composite. In Chapter 9, section 4, we will study a very efficient primality test for Mersenne numbers. The largest known prime numbers are of this form, and their primality was established using this very test.

7. Theorems and proofs

As we have said, the aim of this book is to cover in detail the mathematics at the basis of the RSA cryptosystem. The theoretical backbone was developed by the mathematicians of Ancient Greece, and by Fermat, Euler, and Gauss, and was ready by the end of last century. However, most of the applications were unkown 20 years ago, and some of the results we will mention were proved only a few years back.

Many results in this book won't be new to you. These include the method for computing the greatest common divisor by successive divisions, and the simplest procedures for the factorization of integers into primes. The approach, however, may be new, because we propose to prove every result in the book from first principles, including the procedures to perform computations.

In Ancient Egypt and Mesopotamia, mathematics was a collection of rules of thumb that were used to solve practical problems. It was its association with Greek philosophy that made mathematics the theoretical science it is today. Indeed, the first Greek mathematicians were also famous philosophers like Thales and Pythagoras. The notion that a *mathematical fact* can be *proved* grew out of this interaction with philosophy. After all, a proof is just an argument to show how a certain fact follows from something we know already. And arguing was surely something the Greek philosophers were fond of!

Around 400 B.C., the Greek mathematicians felt the need to spell out, in a more or less systematic way, the hypotheses that served as the foundation of their work. Thus Euclid begins his *Elements* by explicitly stating the definitions and axioms on which his proofs will be based. For example, at the beginning of Book I he defines what he means by point, straight line, plane, surface, and so on. Next, he states the axioms, which he assumes to be self-evident truths. The axioms explain how the previous concepts are interconnected. Then he proceeds to show how far more complex facts about these notions can be reduced, by logical argument, to these axioms. The great advantage of this approach is that it adds solidity to the whole enterprise. By making the foundations more consistent, one can build higher without running the risk of having the whole edifice collapse under its own weight.

A mathematical fact is usually called a *theorem*. This is a Greek word, and it originally meant "spectacle, speculation, theory". The present sense of "a proposition to be proved" is at least as old as Euclid's *Elements*. The statement of a theorem often takes the form of a conditional proposition:

If some *hypothesis* holds, then follows the *conclusion*.

A proof of such a theorem is a logical argument that explains how the conclusion follows from the hypothesis. Here's an example:

Theorem 1. *If a is an even integer, then a^2 is also even.*

In this case, the hypothesis is a *is an even integer*, and the conclusion is a^2 *is an even integer*. Of course, to show that the conclusion follows from the hypothesis we have to use the basic properties of the integers. To make our proofs really watertight, it would be necessary to list all these properties in detail. Needless to say, this would have been impossible in an elementary book like this one. Instead, we will simply pretend that these "basic properties" are the very elementary ones that you already know. These include the rules for adding and multiplying integers, and the fact that between any two given numbers there are only finitely many integers. Let's now use these properties to give a proof of the theorem above.

Proof of theorem 1. The hypothesis says that a is an even integer, which means that it is a multiple of 2; see Chapter 2, section 1. Thus there must exist an integer b such that $a = 2b$. Squaring this last equation, we have

$$a^2 = (2b)^2 = 4b^2 = 2 \cdot (2b^2).$$

Hence a^2 is also a multiple of 2. In other words a^2 is even, which is the conclusion of the theorem.

We saw in theorem 1 that the fact an integer is even *implies* that its square is also even. The *converse* of the conditional statement A *implies* B is the statement B *implies* A. Thus the converse of theorem 1 is *If a^2 is an even integer, then a is an even integer*. Note that the fact a statement is true does not tell us anything about the truth of its converse. For example, the converse of *If an integer is a multiple of* 4, *then it is even* is false: 6 is even but it is not a multiple of 4. When both A *implies* B and B *implies* A are true, we say that A and B are equivalent. This is usually stated in the form A *holds if and only if B holds*. Thus we are led to our second theorem.

Theorem 2. *The integer a is even if and only if a^2 is also even.*

We have already proved that if a is even, then a^2 is even. We must now prove its converse. Before we do that, there is one point of logic we must deal with. Denote the negative of the statement P by *not-P*. For example, if P is "*a is even*", then *not-P* is "*a is not even*". Suppose now that you have two statements P and Q. The proposition *not-Q implies not-P* is called the *contrapositive* of *P implies Q*. Moreover, a statement is true if and only if its contrapositive is true. This only seems odd because it is couched in an unfamiliar language. But consider the following story. Upon being invited to a party, a friend tells you, "my car is broken, but if it is fixed in time, then I'll go to your party". Now if your friend doesn't come to the party, you will conclude that the car wasn't fixed in time—which is just the contrapositive of your friend's statement. Let's go back to the proof of theorem 2.

Proof of theorem 2. We have already seen that if a is even, then a^2 is even; we must now show that if a^2 is even, then a is even. Instead, we will prove its contrapositive, which is, If a is not even, then a^2 is not even. But if an integer is not even, then it is odd. Moreover, an odd integer is always of the form "*even* + 1". Thus, if a is odd, then there must exist another integer b such that $a = 2b + 1$. Squaring both sides of this formula, we have

$$a^2 = (2b+1)^2 = 4b^2 + 4b + 1 = 2 \cdot (2b^2 + 2b) + 1,$$

which is also odd. Thus the contrapositive of the statement that we wanted to prove is true. Since the contrapositive is true if and only if the original statement is true, we have proved that if a^2 is even, then a is even.

We stated theorem 1 in the form If a is even, then a^2 is even. Note that what this really means is that the square of *every* even integer is even. In other words, we are saying that this statement holds for all even numbers. Now consider the statement, Every even number is a multiple of 4. Once again we are claiming that something holds for all even numbers; only this time the claim is false. Why? Because 6, for example, is even, but it is not a multiple of 4. Thus, if someone claims that a property holds for all the elements of a certain set, and we find one element of the set for which it does not hold, then the claim is false. Such an element is called a *counterexample* to the truth of the claim.

The statement of a theorem does not always take the conditional form discussed above. Sometimes it just says that an "object", with certain properties, exists. For example, given a real number x, there is always an integer n such that $n > x$. The most obvious way to prove such a statement is to give an explicit method for finding the object. In the example above, if m is the integer part of x, then $m + 1$ is an integer bigger than x; thus we may take $n = m + 1$. Now, assuming that the decimal expansion of x is known, we can easily use this method to find n. However, one might also prove a statement of this kind without giving any method for constructing the object; this is called a *non-constructive existential proof.* This is not as weird as it seems. Thus we know that any set of 400 persons must always include two with the same birthday, because

$400 > 365$. Although we see that this argument is correct, it does not give us a procedure to find these two persons; thus this is a a non-constructive existential proof.

Most books on number theory make ample use of non-constructive arguments, even when a constructive one is available. This is not just a matter of taste: Constructive proofs are often more awkward to explain than their purely existential counterparts, and mathematicians are as fond of elegance as artists. In this book, however, we will avoid non-constructive proofs as much as possible. The reason is mainly that we are concerned with applications to cryptography. Thus, for example, it is not really satisfactory merely to know that a composite integer has a factor; we must be able to find it.

These brief notes should be enough to allow you to start reading the book. We will have more to say about methods of proof later on, especially in Chapter 2, section 7 and Chapter 5, section 2. But you must realize from the start that the ability to prove theorems needs to be carefully cultivated, and the best way to cultivate it is to practice it often. When Ptolomeus, King of Egypt, asked Euclid if there wasn't an easier way to learn geometry than by reading the *Elements*, the mathematician replied, "There is no royal road to geometry". It was true at the time of Euclid, and it is still true today.

1

Fundamental algorithms

There are two fundamental algorithms: the division algorithm and the Euclidean algorithm. Both were known to the mathematicians of Ancient Greece. Indeed, both appear in Euclid's *Elements*, written around 300 B.C. The division algorithm is used to compute the quotient and the remainder in the division of two integers. The Euclidean algorithm is used to compute the greatest common divisor of two integers. That they are truly fundamental you will realize as you read this book.

1. Algorithms

The sense in which the word *algorithm* is used in this book is defined by the Oxford English Dictionary as follows:

> A process, or set of rules, usually one expressed in algebraic notation, now used especially in computing, machine translation, and linguistics.

In a less roundabout way, we can say that an algorithm is essentially a *recipe* to solve a certain kind of problem.

This suggests that we begin by analyzing a simple recipe in some detail. Suppose we want to make a cake. In a good cookbook, the name of the cake is always followed by the list of the *ingredients* to be used. Then come instructions telling you what to do with the ingredients in order to make the cake. These are things like sift, mix, beat, and bake. Finally, there is the end result: the cake, ready to be eaten.

Every algorithm follows a similar pattern. Thus, when describing an algorithm, we must state its *input* and its *output*. The input corresponds to the ingredients of the recipe. The output is the task we want to get done; in the example above, it is the cake we want to make. The algorithm proper is the set of instructions that tell us what to do to the input in order to get the output.

Suppose we have followed the cake recipe with due care. Of course, upon opening the oven, we expect to find a cake, not a roast beef or a biscuit. We also assume, as we choose the recipe, that it will be possible to have the cake ready after a finite time, preferably a short time. Similarly, we expect of any algorithm that it will produce a result compatible with the announced output. We also expect the algorithm to stop in a finite time, preferably a short time. Of course there are sets of instructions that will run forever. Here is a simple one: Given an integer (the input), add 1 to it, then add 1 to the resulting number, and so on. Since there are infinitely many integers, a program based on these instructions will run forever. Of course, we have no use for a set of instructions like this.

On the other hand, an algorithm may be very slow, but still quite useful. Perhaps no faster algorithm is known; or perhaps the rules are simple, and are used to show that a certain problem has a solution. Of course, it is *not* true that every problem can be solved by following a set of instructions. More surprisingly, there are mathematical problems that cannot be solved algorithmically. Unfortunately, even a brief discussion of this would take us too far from our intended course. For details see Davis 1980.

From an algorithm we may derive a fact, or theorem: Given such and such an input, there is a way (the algorithm) to get a certain output. A theorem is often stated in the form "*If such and such hypotheses hold, then follows the conclusion*". For the theorem associated with an algorithm, the input of the algorithm corresponds to the hypotheses of the theorem, the output to the conclusion.

Don't worry if these comments sound a little vague; we are only setting up the terminology. It will all become clearer when we come to the applications. Summing up, an algorithm is a recipe, a set of instructions, that turns some ingredients (the input) into a certain product (the output). Suppose we are given a set of instructions. How do we decide if it is an algorithm? First of all, we can assume that we are being told what the input and output of this purported algorithm are. Thus, the questions we have to ask are

- When we perform the instructions, do we always arrive at a result after some finite time?
- Is it the expected result?

Having carried the recipe metaphor this far, we should face the fact that one cannot really expect to answer these two questions with respect to a cake recipe. That's because we expect to be able to prove that both questions have an affirmative answer, before we declare that a given sequence of instructions is an algorithm. Of course the word that gives the game away is *proof*. By that we mean a logical reasoning whose starting point is some basic facts, or axioms, that have been previously agreed upon. For most of our algorithms, these axioms will consist of the well-known elementary properties of the integers. Needless to say, one cannot really prove that a cake recipe works in this sense.

The etymology of the word *algorithm* is so peculiar that it deserves our attention. Originally the word was written *algorism*, which comes form the Latinized form of the Arabic *Al-Khowarazmi*. This means "native of Khowarazm", and was the surname of the ninth-century Arab mathematician Abu J'afar Mohamed Ben Musa. It was through his book *Al-jabr wa'l muqabalah* that Arabic numerals became generally known in Europe. Thus *algorism* essentially meant "number", which in Greek is "arithmos". The two words were then "learnedly confused", as the Oxford English Dictionary nicely puts it, giving rise to *algorithm*.

It is not clear how *algorithm* came to mean a "recipe for doing a calculation", but this meaning seems to be quite recent. In English it appears for the first time around 1812. However, one can see that the word was already taking on a more general meaning in the seventeenth century. We have seen that, originally, *algorithm* meant "number", but, by extension, it was also used in the sense of calculation with numbers.

The mathematician and philosopher G. W. Leibniz seems to have been the first to push the use of the word beyond arithmetic. In his first presentation of the calculus, published in 1684, Leibniz refers to the rules of the new calculus as an algorithm. A century later, the word had acquired its present meaning. Gauss, writing in Latin, uses *algoritmus* many times in his *Disquisitiones arithmeticæ* when referring to a set of formulae that constitute a method for finding a solution of some arithmetic problem.

Ben Musa was responsible for at least one more contribution to the terminology of mathematics: The word *algebra* comes from the title of his famous book, mentioned above.

2. Division algorithm

Let us analyze the division algorithm with respect to the scheme set up in the previous section. We are interested in the division of integers, so our task consists of finding the quotient and the remainder in the division of two positive integers. To most of us, "quotients and remainders" bring to mind a picture like this:

$$
\begin{array}{r}
2\,2 \\
5\,4\,\overline{\big)\,1\,2\,3\,4} \\
1\,0\,8 \\
\hline
1\,5\,4 \\
1\,0\,8 \\
\hline
4\,6
\end{array}
$$

In this example we are dividing 1234 by 54 and we have found the quotient to be 22 and the remainder to be 46. In the terminology of section 1, the algorithm has the dividend and the divisor as its input; in the example these are, respectively, 1234 and 54. The output consists of the quotient and the remainder, which, in the example, are 22 and 46, respectively.

In general, the *input* of the division algorithm consists of two positive integers a and b. The algorithm will compute the division of a by b, and the output will be two integers q and r, which are related to a and b as follows:

$$a = bq + r \quad \text{and} \quad 0 \le r < b.$$

Of course q is the quotient, and r is the remainder of the division. This has a simple interpretation that one should always remember. Suppose that we wish to divide a chocolate bar of length a into pieces of length b. The algorithm tells us that we will end up with q pieces of length b, and a smaller piece of length r. It is a good idea to have this in mind even when applying the theorem in a purely mathematical context.

Indeed the chocolate bar is the inspiration for the simplest algorithm for finding q and r, when a and b are given. Although very simple, this algorithm is extremely inefficient.

Division algorithm

Input: positive integers a and b.
Output: non-negative integers q and r such that $a = bq + r$, and $0 \leq r < b$.
Step 1 Begin by setting $Q = 0$ and $R = a$.
Step 2 If $R < b$, write *"the quotient is Q and the remainder is R"* and stop; otherwise, go to step 3.
Step 3 If $R \geq b$, subtract b from R, increase Q by 1, and go back to step 2.

Throughout this book, algorithms will often be presented in the form above. To read the instructions correctly one must abide by some simple conventions. Note that the algorithm makes use of two *variables* Q and R. The variables have been so named because when the algorithm finally stops, they will have taken values that correspond to the quotient and remainder of the division of a by b. In order to compute these numbers we will have to repeat the instructions of steps 2 and 3 several times. This is called a *loop*. At the end of each loop the variables Q and R will have different values. Indeed, that's why they are called variables! The variables change their value in a loop when step 3 is performed. The instruction *subtract b from R* actually means the variable R will now take a new value that is equal to its value at the end of the previous loop minus b. Similarly, *increase Q by* 1 should be taken to mean Q will have a new value that will be equal to the value it had at the end of the previous loop plus 1.

Suppose, for example, that $a > b$. Then, having gone once through step 3, we have $Q = 1$ and $R = a - b$. If $a - b \geq b$, then according to the algorithm, we must apply step 3 one more time. Having done that, we obtain $Q = 2$ and $R = a - 2b$. And so on. Why doesn't it go on forever? In other words, why does the algorithm stop? Note that applying step 3 several times, once for each loop, we get the following sequence of values for the variable R:

Initial value	1st loop	2nd loop	3rd loop	$\dots$
a	$a - b$	$a - 2b$	$a - 3b$	$\dots$

This is a decreasing sequence of integers. Since there are only finitely many integers between a and 0, the sequence must eventually reach a number that is *smaller* than b. When that happens, step 2 instructs us to stop and display the values of R and Q. Thus the algorithm always stops.

We must now see why the end result of the algorithm corresponds to numbers that satisfy the properties specified in the output. First note that if q and r are the numerical values that the variables Q and R have taken when the algorithm stops, then clearly $r = a - bq$ and $r < b$. The first equation immediately gives

$a = bq + r$. We have only to show that $r \geq 0$. We are assuming that the algorithm has stopped in the qth loop, so in the previous loop (which is the $(q-1)$th), we have $a - b(q-1) \geq b$. Otherwise we wouldn't have subtracted b one more time before we stopped. Now, subtracting b from both sides of $a - b(q-1) \geq b$, we get $r = a - bq \geq 0$, which is what we wanted to prove. Hence the algorithm computes whole numbers just like the ones specified in its output.

It is clear that the algorithm presented above is very slow. The number of loops we have to go through is equal to the quotient. Thus we will be in trouble whenever a is a lot bigger than b. The usual method of long division offers a practical way to speed up the process. However, implementing an algorithm based on it is not as straightforward as one might think.

The best way to understand why this is so is to go through the division at the beginning of the section step by step. First of all, we choose the smallest number, formed by the digits of 1234 (from left to right), that is bigger than 54. The number is 123. We now ask, how many times does 54 goes into 123? And it is with this question that the problem lies. If the numbers are small, a couple of trials are enough to lead us to the correct answer. If the numbers are big, we have a problem, because doing it by trial and error may be practically impossible. There is a way out of this dilemma, but to explain it in detail would take us on a long detour; for details see Knuth 1981, section 4.3.

Finally, let us consider a problem of a more practical nature. Most of the results in this book are interesting only when applied to "large integers"—large enough to make division with pencil and paper impracticable. Moreover, we will often have to calculate a dozen divisions when solving a problem. Of course, you can always use a computer algebra system, but a good electronic calculator is enough in many cases. However, when we divide a by b using an electronic calculator, we get the decimal expansion of the fraction a/b—not the quotient and remainder that we need. But consider what we would do if we wanted to find the decimal expansion of a/b using pencil and paper. We would divide the integers until we reached the remainder. Then we would place a decimal point in the quotient, add zeroes to the remainder, and carry on the division. In other words, the *quotient* of the division of a by b is the integer part of the decimal expansion of a/b. Let us call it q. To find the remainder we compute $a - bq$, which is very easy to do with a good electronic calculator.

3. Division theorem

In section 1 we said that to every algorithm there corresponds a theorem. Let's state the theorem that corresponds to the division algorithm.

Division theorem. *Let a and b be positive integers. There exist unique non-negative integers q and r such that*

$$a = bq + r \quad and \quad 0 \leq r < b.$$

Note that the statement says two things about q and r: They always exist, and they are unique. We already know that, given a and b, there exist q and r

as above. We even know how to compute them. But the uniqueness is new. What does it mean to say that q and r are unique? Suppose that we choose two integers a and b, and hand them over to several people, asking them to compute integers q and r so that the relations of the theorem are satisfied. Note that we are only asking people to compute the numbers; we are not telling them any method by which they should do so. The uniqueness of the quotient and the remainder means that *every one of these people will find the same pair of numbers*. In particular, it does not matter which algorithms we choose to compute q and r; they all give the same results. Of course, this is a very useful piece of information.

Let us see why this is true. Suppose we have chosen two positive integers a and b, which we give to two people, Karl and Sofya, and ask them to find quotient and remainder satisfying the conditions stated in the theorem. Karl finds q and r, and Sofya q' and r'. We know only that

$$a = bq + r \quad \text{and} \quad 0 \le r < b$$

and that

$$a = bq' + r' \quad \text{and} \quad 0 \le r' < b.$$

Does this imply that $r = r'$ and $q = q'$? Since r and r' are integers, one of them is greater than or equal to the other one. Say $r \ge r'$. From Karl's equation, we have $r = a - bq$, and from Sofya's we have $r' = a - bq'$. Subtracting them we have

$$r - r' = (a - bq) - (a - bq') = b(q' - q).$$

On the other hand, both r and r' are smaller than b. Since we are also assuming that $r \ge r'$, we have $0 \le r - r' < b$. But $r - r' = b(q' - q)$, so that

$$0 \le b(q' - q) < b.$$

Since b is a positive integer, it can be canceled from this equation. Hence $0 \le q' - q < 1$. But $q' - q$ is an integer, so these inequalities hold if and only if $q' - q = 0$. In other words, $q = q'$, which implies that $r = r'$, and this proves the uniqueness of the quotient and the remainder.

Summing up, we have seen that the division algorithm gives rise to a theorem that says *two* things: The quotient and remainder of the division of two integers always *exist*, and they are *unique*. Many other theorems that will be discussed in this book also state the existence and uniqueness of some property. The most important of them is the *unique factorization theorem* of Chapter 2.

4. The Euclidean algorithm

The Euclidean algorithm is used to compute the greatest common divisor of two integers, and we begin this section by carefully going through the definition of the greatest common divisor.

First of all, we say that an integer b *divides* another integer a if there exists a third integer c such that $a = bc$. In this case we also say that b is a *divisor*, or a *factor*, of a, and also that a is a *multiple* of b. These are only different ways of saying the same thing. The number c in the definition above is called the *co-factor* of b in a. Of course, we find out whether b divides a by computing the remainder of the division and checking that it is zero. The co-factor is then the quotient of the division of a by b.

Suppose that we have two positive integers a and b. The *greatest common divisor* of a and b is the *greatest* positive integer d that divides both a and b. If d is the greatest common divisor of a and b, we write $d = \gcd(a,b)$. If $\gcd(a,b) = 1$, then a and b are said to be *co-prime*.

The definition of the greatest common divisor suggests the following algorithm. Given integers a and b, find all the positive divisors of a and all the positive divisors of b. Check which numbers are common to both sets, and choose the largest of them; it is the greatest common divisor. This is quite simple, but as we will see in the next chapter, it is disastrously inefficient if a or b is big. The problem is that no quick factoring algorithm is known for integers.

Luckily, there is another way to compute the greatest common divisor that is indeed very efficient. It was described by Euclid in propositions 1 and 2 of Book VII of his *Elements*. This is why it is called the *Euclidean algorithm*, even though it may have been known before Euclid.

Let us assume, once again, that a and b are positive integers, and that $a \geq b$. We wish to find the greatest common divisor of a and b. The *Euclidean algorithm* proceeds as follows. First divide a by b; call the remainder of this division r_1. If $r_1 \neq 0$, then divide b by r_1; let r_2 be the remainder of this second division. Similarly, if $r_2 \neq 0$, divide r_1 by r_2 to get a remainder r_3. Thus the ith loop of the algorithm consists essentially of one division in which the dividend is the remainder computed in loop $i-2$, and the divisor is the remainder computed in loop $i-1$. This is repeated until we get a zero remainder; the *smallest non-zero* remainder is the greatest common divisor of a and b.

Let's use the Euclidean algorithm to compute the greatest common divisor of 1234 and 54. The divisions are as follows:

$$1234 = 54 \cdot 22 + 46$$
$$54 = 46 \cdot 1 + 8$$
$$46 = 8 \cdot 5 + 6$$
$$8 = 6 \cdot 1 + 2$$
$$6 = 2 \cdot 3 + 0$$

The last non-zero remainder is 2, so that $\gcd(1234, 54) = 2$. Note that the quotients are not directly used in the computation of the greatest common divisor. Let's now describe the algorithm following the model established in sections 1 and 2.

Euclidean algorithm

Input: positive integers $a \geq b$
Output: the greatest common divisor of a and b

Step 1 Begin by setting $A = a$ and $R = B = b$.

Step 2 Replace the value of R by the remainder of the division of A by B, and go to step 3.

Step 3 If $R = 0$, write *"the greatest common divisor of a and b is B"* and stop; otherwise, go to step 4.

Step 4 Replace the value of A with that of B, and the value of B with that of R; return to step 2.

Therefore, to compute the greatest common divisor we have only to calculate a few divisions. But why does the greatest common divisor coincide with the last non-zero remainder in this sequence of divisions? Indeed, why is zero always a remainder in this sequence? Note that if no zero remainder ever came up, the procedure would never stop.

We begin with the second question. Thus we shall prove first that the algorithm always stops. Suppose that, to find the greatest common divisor of a and b, we compute the divisions below:

$$a = bq_1 + r_1 \quad \text{and} \quad 0 \leq r_1 < b$$
$$b = r_1q_2 + r_2 \quad \text{and} \quad 0 \leq r_2 < r_1$$
$$r_1 = r_2q_3 + r_3 \quad \text{and} \quad 0 \leq r_3 < r_2$$
$$r_2 = r_3q_4 + r_4 \quad \text{and} \quad 0 \leq r_4 < r_3$$
$$\vdots \qquad \cdots \qquad \vdots$$

Forget the left-hand column for the moment. In the right-hand column we have a sequence of remainders. Note that *any given remainder is smaller than the previous one* and also that *all remainders are greater than or equal to zero*. Writing the inequalities one after the other we get

(4.1) $$b > r_1 > r_2 > r_3 > \cdots \geq 0.$$

Since there are only finitely many integers between b and 0, this sequence cannot go on forever. But it will come to an end only if one of the remainders is zero, which means that the algorithm always stops.

We can use the argument of the previous paragraph to get an upper bound on the number of divisions we have to calculate in order to compute the greatest common divisor. Let us go back to (4.1). Each number in the sequence is strictly smaller than the previous one. Hence the largest possible remainder of a certain division equals the previous one minus 1. If that could happen in each division, we would have to compute b divisions to get to a zero remainder. That is clearly the worst possible case. Hence, when we apply the Euclidean algorithm to $a \geq b$, the number of divisions is bounded above by b.

Actually it is not difficult to show that the number of divisions is always smaller than b, unless $b \leq 3$. It is better to state the problem in the following way: What are the smallest *co-prime* integers a and b for which n divisions are required in order to find $\gcd(a, b)$? Note that for the numbers a and b to be as small as possible, the quotients of each division must be as small as possible.

Now, assuming that the divisor is smaller than the dividend, the smallest possible quotient in the division of two integers is clearly 1. Suppose that n divisions are required before we get zero as a remainder. The sequence of remainders is

$$b > r_1 > r_2 > r_3 > r_4 \cdots \geq 0.$$

But we have already seen that, in the worst possible case, the quotients are all 1. Let us now write the divisions, beginning with the *last* one. Since the numbers are co-prime, we have

$$r_{n-1} = 1$$
$$r_{n-3} = r_{n-2} \cdot 1 + 1$$
$$r_{n-4} = r_{n-3} \cdot 1 + r_{n-2}$$
$$\cdots$$
$$a = b \cdot 1 + r_1$$

For $n = 10$, we get the following sequence of remainders (written in decreasing order):

$$34, \; 21, \; 13, \; 8, \; 5, \; 3, \; 2, \; 1, \; 1, \; 0.$$

Hence the smallest co-prime integers a and b for which 10 divisions are necessary in order to calculate $\gcd(a, b)$ are $a = 34$ and $b = 21$. Note that although $b = 21$ is the smallest possible value, it is bigger than $n = 10$. The sequence above is the well-known *Fibonacci sequence*. It will reappear in exercise 6.

5. Proof of the Euclidean algorithm

We have seen that the algorithm always stops. Indeed, it will never have to calculate more divisions than the smallest of the two numbers whose greatest common divisor we want to find. But why is the last non-zero remainder exactly the greatest common divisor? To understand this we need an auxiliary result, what mathematicians call a *lemma*. The word comes from the Ancient Greek, and it means "something that one assumes" in order to prove a theorem.

Lemma. *Let a and b be positive integers. Suppose that there exist integers g and s such that $a = bg + s$. Then $\gcd(a, b) = \gcd(b, s)$.*

We must show that the result stated in the lemma is true. But before we do that, let us use the lemma to prove that the last non-zero remainder in the Euclidean algorithm is indeed equal to the greatest common divisor. Applying the algorithm to integers $a \geq b > 0$, and assuming that the remainder is zero after n divisions, we have

(5.1)
$$a = bq_1 + r_1 \quad \text{and} \quad 0 \leq r_1 < b$$
$$b = r_1 q_2 + r_2 \quad \text{and} \quad 0 \leq r_2 < r_1$$

$$r_1 = r_2q_3 + r_3 \quad \text{and} \quad 0 \le r_3 < r_2$$
$$r_2 = r_3q_4 + r_4 \quad \text{and} \quad 0 \le r_4 < r_3$$

$$\vdots \qquad \cdots \qquad \vdots$$

$$r_{n-4} = r_{n-3}q_{n-2} + r_{n-2} \quad \text{and} \quad 0 \le r_{n-2} < r_{n-3}$$
$$r_{n-3} = r_{n-2}q_{n-1} + r_{n-1} \quad \text{and} \quad 0 \le r_{n-1} < r_{n-2}$$
$$r_{n-2} = r_{n-1}q_n \quad \text{and} \quad r_n = 0$$

This time we can ignore the right-hand column and consider only what happens in the left-hand column. The last division tells us that r_{n-1} divides r_{n-2}. Hence the greatest common divisor of these two numbers is r_{n-1}. In other words, $\gcd(r_{n-2}, r_{n-1}) = r_{n-1}$.

Now the lemma comes into action. Applying the lemma to the second-to-last division, we conclude that

$$\gcd(r_{n-3}, r_{n-2}) = \gcd(r_{n-2}, r_{n-1}),$$

which, we have seen, is equal to r_{n-1}. Applying the lemma again, this time to the antepenultimate division, we have

$$\gcd(r_{n-4}, r_{n-3}) = \gcd(r_{n-3}, r_{n-2}),$$

which we know equals r_{n-1}. Going on like this all the way to the top of the column, we get $\gcd(a, b) = r_{n-1}$, which is what we wanted to prove.

The proof will be complete once we have proved the lemma. Recall that it says, assuming a, b, g and s to be related by $a = bg + s$, then it should follow that $\gcd(a, b) = \gcd(b, s)$. It will be easier to explain the proof if we write

$$d_1 = \gcd(a, b) \quad \text{and} \quad d_2 = \gcd(b, s).$$

Of course we have not done anything; we have only given special names to the greatest common divisors of a and b and of b and s. What we want to prove is that $d_1 = d_2$. We will do this in two steps. First we show that $d_1 \le d_2$, and then that $d_2 \le d_1$. The equality of d_1 and d_2 follows immediately from these two inequalities.

Let us show that $d_1 \le d_2$; the other inequality is proved by a similar argument and will be left as an exercise. Recall that $d_1 = \gcd(a, b)$. Thus d_1 divides a, and d_1 divides b. This means that there exist integers u and v such that

$$a = d_1u \quad \text{and} \quad b = d_1v.$$

Replacing a and b in $a = bg + s$ respectively by d_1u and d_2v, we get $d_1u = d_1vg + s$. In other words,

$$s = d_1u - d_1vg = d_1(u - vg).$$

But this means that d_1 divides s.

Summing up, we have by hypothesis that $d_1 = \gcd(a, b)$, so d_1 divides b. But the calculations above show that d_1 also divides s. Hence d_1 is a common divisor of b and s. However, d_2 is the *greatest* common divisor of b and s. Therefore $d_1 \leq d_2$, which is the inequality we wanted to prove.

Note that the proof makes essential use of the relation $a = bg + s$, which is analogous to the relation in the division theorem. However, we need not know that s is less than b; indeed, it need not even be positive. Hence the fact that the remainder is less than the divisor is not used to show that the last non-zero remainder is the greatest common divisor, but only to show that the algorithm stops.

6. Extended Euclidean algorithm

There is a still more powerful version of the Euclidean algorithm than the one described in the previous section. In this case, powerful does not mean faster. It means that the greatest common divisor is only one of the elements of the output. Suppose, once again, that a and b are positive integers, and let d be their greatest common divisor. The *extended Euclidean algorithm* will calculate d, and also two integers α and β, such that

$$(6.1) \qquad \alpha \cdot a + \beta \cdot b = d.$$

Note that (except in some trivial cases), we will always have that, if α is positive, then β is negative, and vice versa.

The best way to compute these integers is to add some extra calculations to the traditional Euclidean algorithm, so that d, α, and β will be found at the same time. This explains why the resulting procedure is known as the *extended* Euclidean algorithm. The version of the algorithm we present here is the creation of D. E. Knuth, author of the famous book *The Art of Computer Programming*. The algorithm can be found in volume 2 of the series; see Knuth 1981, section 4.5.2, algorithm X.

Recall that the Euclidean algorithm proceeds by computing a sequence of divisions. The greatest common multiple is the last non-zero remainder in this sequence. Thus we have to find a way of writing the last non-zero remainder as in (6.1).

The idea behind Knuth's algorithm is that we should not wait to arrive at the last remainder; instead we should find a way of writing each remainder, from first to last, in the required way. This apparently means that we have to do a lot of extra unnecessary work. That's not really true, as we will see later on in this section.

Suppose that, in order to compute the greatest common divisor of a and b, we go through the sequence of divisions (5.1). Let us write them here together with the special formulae for the remainders that we expect to find.

$$(6.2) \qquad \begin{aligned} a &= bq_1 + r_1 \quad \text{and} \quad r_1 = ax_1 + by_1 \\ b &= r_1q_2 + r_2 \quad \text{and} \quad r_2 = ax_2 + by_2 \end{aligned}$$

$$r_1 = r_2 q_3 + r_3 \quad \text{and} \quad r_3 = ax_3 + by_3$$
$$r_2 = r_3 q_4 + r_4 \quad \text{and} \quad r_4 = ax_4 + by_4$$

$$\vdots \quad \cdots \quad \vdots$$

$$r_{n-3} = r_{n-2} q_{n-1} + r_{n-1} \quad \text{and} \quad r_{n-1} = ax_{n-1} + by_{n-1}$$
$$r_{n-2} = r_{n-1} q_n \quad \text{and} \quad r_n = 0.$$

The numbers $x_1, \ldots, x_{n-1}$ and $y_1, \ldots, y_{n-1}$ are the integers that we want to determine. It is convenient to summarize the information we require from (6.2) in a table.

remainders	quotients	x	y
a	*	x_{-1}	y_{-1}
b	*	x_0	y_0
r_1	q_1	x_1	y_1
r_2	q_2	x_2	y_2
r_3	q_3	x_3	y_3
$\vdots$	$\vdots$	$\vdots$	$\vdots$
r_{n-2}	q_{n-2}	x_{n-2}	y_{n-2}
r_{n-1}	q_{n-1}	x_{n-1}	y_{n-1}

The first thing to note is that the table begins with two rows that "legally" should not be there. Indeed the numbers that appear in the first column of these rows are not really remainders of any divisions. Let us call these the (-1)th and the 0th rows, thus emphasizing their "outlaw" character. Soon we will see why they are necessary.

What, then, do we want to do? We want to find out how to fill in columns x and y. Suppose, for a moment, that we have received the table filled in up to a certain row, say the $(j-1)$th row. The first thing we have to do in order to fill in the jth-row is to divide r_{j-2} by r_{j-1}. This gives r_j and q_j, which fill in the first and second positions of this row. Let us not forget that $r_{j-2} = r_{j-1} q_j + r_j$ and $0 \leq r_j < r_{j-1}$. Thus,

$$(6.3) \qquad\qquad r_j = r_{j-2} - r_{j-1} q_j.$$

Now, in rows $j-1$ and $j-2$, we find the values of $x_{j-2}, x_{j-1}, y_{j-2}$ and y_{j-1}. So we can write

$$r_{j-2} = ax_{j-2} + by_{j-2} \quad \text{and} \quad r_{j-1} = ax_{j-1} + by_{j-1}.$$

Inserting these values in (6.3), we obtain

$$\begin{aligned} r_j &= (ax_{j-2} + by_{j-2}) - (ax_{j-1} + by_{j-1})q_j \\ &= a(x_{j-2} - q_j x_{j-1}) + b(y_{j-2} - q_j y_{j-1}). \end{aligned}$$

Hence

$$x_j = x_{j-2} - q_j x_{j-1} \quad \text{and} \quad y_j = y_{j-2} - q_j y_{j-1}.$$

Note that to compute x_j and y_j we have only used the quotient q_j and data from the two rows that immediately precede row j. This explains why Knuth's algorithm is so efficient. In order to fill in a certain row we need only the two rows that immediately precede it; all the other rows can now be deleted from the memory of the computer.

Thus we have a recursive process; all we have to do is find out how to give it a push. This is why we need the two "illegal" rows that we introduced at the beginning of the table. The reason to introduce them is that it is very easy to compute the values of x and y for these rows. Interpreting x and y just as we did for the other rows, we must have that

$$a = ax_{-1} + by_{-1} \quad \text{and} \quad b = ax_0 + by_0.$$

This suggests that we choose

$$x_{-1} = 1, y_{-1} = 0, x_0 = 0 \quad \text{and} \quad y_0 = 1,$$

which is enough to give the initial push to the procedure.

Having gone through the divisions, we know that $\gcd(a, b) = r_{n-1}$, and we have computed integers x_{n-1} and y_{n-1} such that

$$d = r_{n-1} = ax_{n-1} + by_{n-1}.$$

Hence $\alpha = x_{n-1}$ and $\beta = y_{n-1}$. Note that if we know α and $d = r_{n-1}$, then we can find β by the formula

$$\beta = (d - a\alpha)/b.$$

Therefore we need only compute the first three rows of the table.

Here is a numerical example. If $a = 1234$ and $b = 54$, then the (full) table will be as follows.

remainders	quotients	x	y
1234	*	1	0
54	*	0	1
46	22	$1 - 22 \cdot 0 = 1$	$0 - 22 \cdot 1 = -22$
8	1	$0 - 1 \cdot 1 = -1$	$1 - 1 \cdot (-22) = 23$
6	5	$1 - 5 \cdot (-1) = 6$	$-22 - 5 \cdot 23 = -137$
2	1	$-1 - 1 \cdot 6 = -7$	$23 - 1 \cdot (-137) = 160$
0	3	*	*

Hence $\alpha = -7$, $\beta = 160$, and

$$(-7) \cdot 1234 + 160 \cdot 54 = 2.$$

It is time to find out why the algorithm works, and why it stops. As the name suggests, this is the Euclidean algorithm of the previous section with some extra instructions added for the calculation of x and y. Thus it stops, and the greatest common divisor is part of its output, because this is true of the Euclidean algorithm. Moreover, the numbers in the columns x and y of each row satisfy an equation like (6.1), with d replaced by the remainder of that row. In particular, (6.1) holds if we choose for α and β the numbers in the columns x and y of the row corresponding to the last non-zero remainder. The *extended Euclidean algorithm* gives rise to the following theorem.

Theorem. *Let a and b be positive integers, and let d be the greatest common divisor of a and b. There exist integers α and β such that*

$$\alpha \cdot a + \beta \cdot b = d.$$

Note that the numbers α and β are not unique. Indeed, there are infinitely many possible choices of integers α and β for which (6.1) is satisfied. For example, let k be an integer, and suppose that α and β are such that $\alpha \cdot a + \beta \cdot b = d$. Then

$$(\alpha + kb) \cdot a + (\beta - ka) \cdot b = d,$$

as one immediately checks.

Having gone through all this effort in order to calculate α and β, it is reasonable to ask, What are these numbers good for? The best way to find out the answer is to keep reading the book. Many of the most important results of later chapters will depend on the knowledge of these numbers, and this includes the choice of keys of the RSA cryptosystem in Chapter 11.

7. Exercises

1. For each pair a, b of integers given below, compute the greatest common divisor and find integers α, β such that $\gcd(a, b) = \alpha \cdot a + \beta \cdot b$.

    (1) 14 and 35
    (2) 252 and 180
    (3) 6643 and 2873
    (4) 272,828,282 and 3242

2. Let n be an integer greater than 1. Show that

    (1) $\gcd(n, 2n + 1) = 1$.
    (2) $\gcd(2n + 1, 3n + 1) = 1$.
    (3) $\gcd(n! + 1, (n + 1)! + 1) = 1$.

3. Show that if a, b and $n > 0$ are integers, then

$$b^n - a^n = (b - a)(b^{n-1} + b^{n-2}a + b^{n-3}a^2 + \cdots + ba^{n-2} + a^{n-1}).$$

4. Let $n > m$ be positive integers and let r be the remainder of the division of n by m.

    (1) Show that the remainder of the division of $2^n - 1$ by $2^m - 1$ is $2^r - 1$.
    (2) Show that if the quotient of n by m is even, then the remainder of the division of $2^n + 1$ by $2^m + 1$ is $2^r + 1$.

Hint: Use exercise 3 to compute the quotient in each case; the result follows from the uniqueness of the remainder.

5. Let $n > m$ be positive integers. Use exercise 4 to compute $\gcd(2^{2^n} + 1, 2^{2^m} + 1)$. The result of this exercise is used in exercise 8 of Chapter 3.

6. In the Fibonacci sequence $1, 1, 2, 3, 5, 8, 13 \ldots$, each number is the sum of the two that come immediately before it. Denoting by f_n the nth number in the sequence, we have that

$$f_n = f_{n-1} + f_{n-2},$$

and that $f_0 = f_1 = 1$.

  (1) Show that the greatest common divisor of two consecutive terms of the Fibonacci sequence is 1.
  (2) How many divisions are required to compute $\gcd(f_n, f_{n-1})$?

7. In this exercise we describe a method that can be used to find the integer solutions of the equation $ax + by = c$, where $a, b, c \in \mathbb{Z}$. In other words, we want either to find integers x and y that satisfy the equation or to prove that they do not exist. Let $d = \gcd(a, b)$. Then, $a = da'$ and $b = db'$, for some integers a' and b'. Hence,

$$c = ax + by = d(a'x + b'y).$$

Show that if the equation has an integer solution, then d divides c.

If that is the case, let $c = dc'$ and consider the *reduced equation* $a'x + b'y = c'$. Show that any solution of the original equation is also a solution of the reduced equation, and vice versa.

Hence, to find the solutions of the original equation it is enough to solve the reduced equation. In order to do that we use the extended Euclidean algorithm to compute integers α and β such that $\alpha \cdot a + \beta \cdot b = 1$. Show that, in this case, the reduced equation has solutions $x = c'\alpha$ and $y = c'\beta$.

8. Use the method of exercise 7 to write a program to solve the equation $ax + by = c$ in integers. The input will be the coefficients a, b, and c. The output will be either an integer solution of the equation or a message saying that such a solution does not exist. Thus the program will consist essentially of an implementation of the extended Euclidean algorithm.

9. The purpose of this exercise is to find out experimentally what proportion of randomly generated pairs of integers is composed of co-prime pairs. The input of the program will be a positive number m, the total number of random pairs to be generated. The program will apply the Euclidean algorithm to these pairs, find their greatest common divisors, and count how many of these equal 1. The output will be the quotient

$$\frac{\text{total number of co-prime pairs}}{m}.$$

This quotient gives a measure of the probability that a randomly chosen pair of integers is co-prime. It is necessary to apply the program to large values of m in order to get a good approximation of the probability. Run the program ten times with $m = 10^5$. What were the values you obtained for the output? One can show, from theoretical considerations, that the correct probability is $6/\pi^2$; see Knuth 1981, section 4.5.2, Theorem D. How does your experimental value compare with this number?

2

Unique factorization

"Divide and conquer" is a very common strategy in science. For example, any substance can be broken into its constituent parts, the atoms. Moreover, if the properties of the atoms are known in sufficient detail, that tells us a lot about the substances that are made of them.

Something similar happens with the integers. In this case the role of atoms is played by the *prime numbers*, and every integer can be written as a product of primes. This decomposition is a crucial ingredient in the proof of many properties of the integers. However, it is not always easy to compute the decomposition of a given integer. If the number is large, factorization can be a very time-consuming process, making heavy demands of computer power.

1. Unique factorization theorem

Let us begin by carefully defining the main characters. An integer p is said to be *prime* if $p \neq \pm 1$ and its only divisors are ± 1 and $\pm p$. Hence 2, 3, 5, and -7 are prime numbers, but $45 = 5 \cdot 9$ is not. Notwithstanding the definition, we will use the word "prime" as shorthand for "positive prime" almost everywhere in the book. An integer, different from ± 1, that is not prime is said to be *composite*. Thus, if n is composite, there exist integers a and b such that $1 < a, b < n$ and $n = ab$. Hence 45 is composite.

Note that the numbers ± 1 are neither composite nor prime. They belong to a third category: They are the only integers to have a multiplicative inverse. At the end of this section we will be able to explain in a more convincing way why these numbers should be left out of the set of primes.

Unique factorization theorem. *An integer $n \geq 2$ can be uniquely written in the form*

$$n = p_1^{e_1} \dots p_k^{e_k},$$

where $1 < p_1 < p_2 < p_3 < \cdots < p_k$ are prime numbers, and $e_1, \dots, e_k$ are positive integers.

This theorem is so important that it is sometimes called the *fundamental theorem of arithmetic*. It was first stated in this form by C. F. Gauss in section 16 of his *Disquisitiones arithmeticæ*. That, however, did not stop earlier mathematicians from using the theorem implicitly. Indeed, as Hardy and Wright write in their book on number theory, "Gauss was the first to develop arithmetic as a systematic science"; see Hardy and Wright 1994, p. 10.

The exponents $e_1, \ldots, e_k$ are called the *multiplicities* of the primes in the factorization of n. In other words, the multiplicity of p_1 in the factorization of n is the *largest* integer e_1 such that $p_1^{e_1}$ divides n. Note also that n has k *distinct* prime factors, but the *total* number of prime factors is $e_1 + \cdots + e_k$.

The statement of the theorem says two different things. First, every integer can be written as the product of powers of prime numbers. Second, there is only one possible choice of primes and exponents for the decomposition of a given integer. Thus, we have two things to prove: that the factorization exists, and that it is unique. They will be proved separately. As we will see, it is easier to prove that the factorization exists. The uniqueness is far more subtle.

Having stated the unique factorization theorem, we are in a better position to explain why ± 1 should not be counted among the primes. If we had defined primes so as to include these numbers, then the factorization of integers into primes would no longer be unique. Indeed, if 1 is prime, then 2 and $1^2 \cdot 2$ are two different factorizations of 2 in powers of primes. Playing the same game with higher powers of 1 (or -1), we would have infinitely many different factorizations for every integer. Thus, to avoid these pseudo-factorizations (infinite in number, but quite worthless) we exclude ± 1 from the definition of prime number.

Finally, we come across an interesting etymological question: How did prime numbers get their name? The mathematicians of Ancient Greece classified whole numbers as *primary* or *indecomposable*, and *secondary* or *composite*. The Greek for *primary* was translated as "primus" in Latin, which in turn became *prime* in English.

2. Existence of the factorization

We show in this section that, given an integer $n \geq 2$, it can be written as a product of primes. We do this by describing an algorithm that has as its input an integer $n \geq 2$, and as its output the prime factors of n and their multiplicities. As a preliminary stage to this algorithm, we consider another one whose output is a single prime factor of n.

The simplest algorithm to find prime factors is the following. Suppose the input is an integer $n \geq 2$. Try to divide n by the integers from 2 to $n-1$. If one of these integers divides n, then n is composite, and we have found its smallest factor. Otherwise, n is prime. Moreover, if n is composite, the factor we found must be prime.

Let's see why this last statement is true. Let f be an integer such that $2 \leq f \leq n - 1$. Suppose that f is the *smallest* factor of n, and let $f' > 1$ be a factor of f. By the definition of divisibility, there exist integers a and b such that

$$ n = f \cdot a \quad \text{and} \quad f = f' \cdot b. $$

Hence $n = f' \cdot ab$. Thus f' is also a factor of n. Since f is the *smallest* factor of n, we must have $f \leq f'$. But f' is a factor of f, so $f' \leq f$. From these inequalities it follows that $f = f'$. Therefore, we have proved that if $f' \neq 1$ divides f, then $f' = f$. In other words, f is prime.

There is something else we ought to note before we describe the algorithm in detail. We have seen that all the algorithm does is search for factors among the positive integers. How far do we have to carry this search? It is obvious that we need not go beyond $n - 1$; an integer cannot have a factor bigger than itself. But we can do better. Indeed, it is not necessary to look for factors bigger than $\sqrt{n}$. Once again this depends on the fact that the algorithm actually finds the *smallest* factor of n, greater than 1. Thus, all we have to show is that if n is *composite* and $f > 1$ is the *smallest* factor of n, then $f \leq \sqrt{n}$.

This last claim is easily checked. Let a be the co-factor of f in n; thus $n = fa$. Since $f > 1$ is the smallest factor of n, we have $f \leq a$. Note that this would *not* be true if n were not composite. Now $a = n/f$, and so $f \leq n/f$. But this implies that $f^2 \leq n$. In other words, $f \leq \sqrt{n}$, which is what we wanted to prove.

The discussion can be summed up as follows. The algorithm consists of a search for factors, beginning with 2 and moving through the integers up to $\sqrt{n}$. If n is composite, its smallest factor greater than or equal to 2 will be found. This factor is necessarily prime. If no factor is found in this search, then n itself is prime.

There is one final point of a practical nature. Denote by $[\alpha]$ the integer part of a real number α. In other words, $[\alpha]$ is the greatest integer smaller than or equal to α. Thus $[\pi] = 3$ and $[\sqrt{2}] = 1$. Note that if r is an integer, and $r \leq \alpha$, then $r \leq [\alpha]$. Thus, to use the factorization algorithm described above we need only know $[\sqrt{n}]$. A procedure for doing this can be found in section 1 of the Appendix.

Next we will write the factorization algorithm according to the canon of Chapter 1. In order to do this with a minimum of fuss, we will assume that we are using a computer that can determine the integer part of $\sqrt{n}$.

Factorization by trial division

Input: positive integer n
Output: positive integer $f > 1$, which is the smallest prime factor of n, or a message stating that n is prime

Step 1 Begin by setting $F = 2$.
Step 2 If n/F is an integer, write "*F is a factor of n*" and stop; otherwise go to step 3.
Step 3 Increase F by one, and go to step 4.
Step 4 If $F \geq [\sqrt{n}]$ write "*n is prime*" and stop; otherwise return to step 2.

Given an integer $n > 2$, we have found a method to determine whether n is prime, which also finds a factor of n if it is composite. Of course, if n is prime, we have already found its factorization. However, if n is composite, we wish to find all its prime factors with their respective multiplicities. In order to do this it is enough to apply the above algorithm several times.

Suppose that, having applied the algorithm to n, we found a factor q_1. Hence q_1 is the smallest prime factor of n. Next we apply the algorithm to the co-factor n/q_1. Suppose that n/q_1 is composite and that its smallest prime factor is q_2. Clearly $q_2 \geq q_1$. But note that they could be equal. That would happen if q_1^2 divided n. Carrying on in the same fashion, we would have to apply the algorithm to $n/(q_1 q_2)$, and so on. This method generates an increasing sequence of primes,

$$q_1 \leq q_2 \leq q_3 \leq \cdots \leq q_s,$$

each one of which is a factor of n. To this sequence there corresponds another one, the sequence of co-factors

$$\frac{n}{q_1} > \frac{n}{q_1 q_2} > \frac{n}{q_1 q_2 q_3} > \ldots.$$

Note that this is a strictly *decreasing sequence* of positive integers, each of which corresponds to an application of the trial division algorithm to n. Since there are only finitely many positive integers smaller than n, the complete factorization of n will be found after a finite number of steps. It is not difficult to check that the last number in the sequence of co-factors is always 1, which gives a simple clue that it is time to stop.

Now suppose we want to write the factorization of n in the form that appears in the statement of the unique factorization theorem. We have the prime factors; all we need are their multiplicities. To find them, it is enough to count how many times each prime appears in the sequence of primes above. Of course, it is better to count the primes as we carry on the process.

Let us see how this works in an example. Suppose we want to find the factorization of $n = 450$. Applying the trial division algorithm, we find that the smallest factor of 450 is 2. Applying the same algorithm again, this time to the co-factor $450/2 = 225$, we find the factor 3. Thus, 3 is a factor of 225, and so also a factor of 450. Next, apply the algorithm to $225/3 = 75$. Once again the smallest factor is 3. Hence, 3^2 divides 450. Two more applications of the trial division algorithm allow us to show that 5^2 divides 25, and has 1 as its co-factor. Hence, we have found the complete factorization of 450, which is $450 = 2 \cdot 3^2 \cdot 5^2$.

3. Efficiency of the trial division algorithm

The algorithms described in the previous section are easy to understand and to program, but they are very inefficient. Since the algorithm used to find the complete factorization of an integer works by repeated application of the trial division algorithm, it is enough to discuss the efficiency of the latter. We illustrate the point with a simple, but quite dramatic, calculation.

If we apply the trial division algorithm to an integer $n > 2$, then the worst possible case occurs when n is prime. In that case the algorithm will repeat the loop $[\sqrt{n}]$ times before it stops. So, to keep the calculations simple, let us assume that we have a prime number n with 100 digits or more. How long would it take to certify that n is prime with the trial division algorithm?

We are assuming that $n > 10^{100}$, so $\sqrt{n} > 10^{50}$. Thus, we have to repeat the loop in the algorithm at least 10^{50} times. In order to find out how long it would take to do this, let us suppose that our computer calculates 10^{10} divisions per second. Note that in doing this, we are pretending that the only operation the computer has to perform in a loop is division. Of course, this is far from true; but let that pass. Dividing the numbers, we find that the computer would take 10^{40} seconds to prove that n is prime. A simple calculation shows that this is approximately equal to 10^{31} years. This is far too large a number by everyday standards. To put it in perspective, we might compare it with the age of the universe. Present calculations indicate that the Big Bang occurred some $2 \cdot 10^{11}$ years ago. Surely no additional comment is required; the numbers make the message clear.

Does that mean this factorization algorithm is useless? Of course not. Perhaps the number we wish to factor has a small prime factor, say, smaller than 10^6. In that case the trial division algorithm would find it quickly. On the other hand, if we have reasons to think that a given large number is prime, then the trial division algorithm is not the best way to go about proving it.

There are many other algorithms for the factorization of integers, which are more or less efficient depending on the kind of integer one wants to factor. Thus, the algorithm of section 2 is quite good for integers with small factors. In the next section we will study Fermat's algorithm, which is most efficient for integers n that have a (not necessarily prime) factor not much bigger than $\sqrt{n}$.

One should not forget that there is no algorithm that is efficient for the factorization of any randomly chosen integer. Otherwise the RSA cryptosystem would not be truly secure. What is not clear is whether such an algorithm does not exist, or whether we just haven't been clever enough to discover one.

4. Fermat's factorization algorithm

The algorithm of section 2 is efficient only if the integer n that we wish to factor is divisible by a small prime. How "small" will depend on the computer. In this section we study an algorithm that is efficient when n has a (not necessarily prime) factor not much bigger than $\sqrt{n}$. This algorithm was conceived by Fermat, and it is a lot more ingenious than the trial division algorithm.

To start with, suppose that n is odd. If it were even, 2 would be one of its factors. The idea behind the algorithm is to try to find non-negative integers x and y, such that $n = x^2 - y^2$. If we assume that these numbers have been found, then

$$n = x^2 - y^2 = (x - y)(x + y).$$

Hence $x - y$ and $x + y$ are factors of n.

In order to describe the algorithm with a minimum of fuss, we will assume that we are using a computer that can determine the integer part of $\sqrt{n}$.

The easiest case of Fermat's algorithm occurs when n is a perfect square. That is, when $n = r^2$ for some integer r. Then r is a factor of n. Note that, in the notation above, in this case we have $x = r$ and $y = 0$.

On the other hand, if $y > 0$, then

$$x = \sqrt{n + y^2} > \sqrt{n}.$$

This suggests the following strategy for finding x and y.

Fermat's factorization algorithm

Input: a positive odd integer n

Output: a factor of n, or a message stating that n is prime

Step 1 Begin with $x = [\sqrt{n}]$. If $n = x^2$, then x is a factor of n, and we can stop; otherwise increase x by 1 and go to step 2.

Step 2 If $x = (n + 1)/2$, then n is prime and we can stop; otherwise compute $y = \sqrt{x^2 - n}$.

Step 3 If y is an integer (that is, if $[y]^2 = x^2 - n$) then n has factors $x + y$ and $x - y$ and we can stop; otherwise increase x by 1 and go to step 2.

The algorithm is very easy to use, as the following example shows. Let $n = 1342127$ be the number we want to factor. The variable x is initialized as the integer part of $\sqrt{n}$. In this example this gives $x = 1158$. However,

$$x^2 = 1158^2 = 1,340,964 < 1,342,127.$$

Thus we must increase x by 1. We will keep on doing so until $\sqrt{x^2 - n}$ is an integer, or $x = (n + 1)/2$, whichever comes first. Note that, in this example, $(n + 1)/2 = 67,1064$. The values of the variables x and y at the various loops are easily summed up in a table.

x	$\sqrt{x^2 - n}$
1159	33.97
1160	58.93
1161	76.11
1162	90.09
1163	102.18
1164	113

Thus, we have found an integer in the sixth loop. Hence $x = 1164$ and $y = 113$ are the numbers we want. The corresponding factors are

$$x + y = 1277 \quad \text{and} \quad x - y = 1051.$$

5. Proof of Fermat's algorithm

We must now prove that Fermat's algorithm works, and that it always stops. In doing so, it is convenient to consider separately the behavior of the algorithm when the input n is composite, and when it is prime. In the first case we must show that there exists an integer x such that $[\sqrt{n}] \le x < (n+1)/2$, and $\sqrt{x^2 - n}$ is an integer. This means that if n is composite, then the algorithm always finds a factor smaller than n before x becomes equal to $(n+1)/2$. Now if n is prime, we have to check that $\sqrt{x^2 - n}$ is never an integer for $x < (n+1)/2$.

Suppose that n can be factored in the form $n = ab$, where $a \le b$. We wish to find integers x and y such that $n = x^2 - y^2$. In other words,

$$n = ab = (x - y)(x + y) = x^2 - y^2.$$

Since $x - y \le x + y$, we can try the choice $a = x - y$ and $b = x + y$. Solving this system of equations in two unknowns, we get

$$x = \frac{a+b}{2} \quad \text{and} \quad y = \frac{b-a}{2}.$$

Indeed, a simple calculation shows that

(5.1)
$$\left(\frac{b+a}{2}\right)^2 - \left(\frac{b-a}{2}\right)^2 = ab = n.$$

Note that since x and y must be integers, both $(b + a)$ and $(b - a)$ need to be even numbers. This is why n has to be odd. For a and b are factors of n, so they must also be odd; thus $a + b$ and $b - a$ are even. If n is even, the algorithm may not work properly. For example, if $n = 2k$ and k is odd, the algorithm will never stop.

If n is prime, the only possible choices for a and b are $a = 1$ and $b = n$. Thus $x = (n + 1)/2$, and this is the smallest x for which $\sqrt{x^2 - n}$ is an integer. We must now consider what happens when n is composite. If $a = b$, then the algorithm finds a factor in step 1. Thus, we may assume that n is composite and that it is *not* a perfect square. In other words, $1 < a < b < n$. We claim that in this case the algorithm stops because

(5.2)
$$[\sqrt{n}] < \frac{a+b}{2} < \frac{n+1}{2}.$$

Let us prove the inequalities first.

The inequality on the right says that $a + b < n + 1$. Replacing n by ab, and subtracting $b + 1$ from both sides, we get $a - 1 < ab - b$. But $a > 1$, so we can cancel $a - 1$ from both sides of the inequality. Having done so, we find that $1 < b$. This argument shows that the inequality $1 < b$ is equivalent to $a + b < n + 1$. Since $1 < a < b$ holds by hypothesis, we have shown that $(a + b)/2 < (n + 1)/2$.

Let us now consider the inequality on the left-hand side. First note that since $[\sqrt{n}] \le \sqrt{n}$, it is enough to prove that $\sqrt{n} \le (a + b)/2$. Clearly this last inequality holds if and only if $n \le (a + b)^2/4$. However, by (5.1)

$$\frac{(b+a)^2}{4} - n = \frac{(b-a)^2}{4},$$

which is always non-negative. Hence we have proved that $(a + b)^2/4 - n \ge 0$, which is equivalent to the inequality we began with.

Let's go back to the algorithm. Recall that the variable x is initialized with the value $[\sqrt{n}]$, and then increased by 1 at each loop. Hence it follows from (5.2) that, if n is composite, the algorithm will reach $(a+b)/2$ before it arrives at $(n+1)/2$. However, when $x = (a+b)/2$, we get

$$y^2 = \left(\frac{a+b}{2}\right)^2 - n = \left(\frac{b-a}{2}\right)^2 .$$

Thus, having reached this value of x, the algorithm stops, and the output will be the factors a and b. Therefore, if n is composite, the algorithm always stops for some $x < (n+1)/2$, having computed two factors of n.

Note that given a composite integer n, it may be possible to factor n in many different ways in the form $n = ab$, with $1 < a < b < n$. Which of these is the one Fermat's algorithm will find? The algorithm begins its search for x at $[\sqrt{n}]$, increasing x at each loop. Thus the factors a and b that the algorithm will find are those for which

$$\frac{a+b}{2} - [\sqrt{n}]$$

is as small as possible.

This algorithm has something very important to tell us about the RSA cryptosystem. Recall that the security of the RSA depends on the difficulty of factoring an integer n, which is the product of two primes. If we can factor n, we can break the code. The trial division algorithm might give the illusion that by choosing big primes we can make sure that n cannot be easily factored. But this is *not* true. If the primes are big but their difference is small, then n will be very easily factored by Fermat's algorithm. We shall return to this question in Chapter 11.

6. A fundamental property of primes

In order to prove that the prime factorization of an integer is unique, we need a fundamental property of prime numbers. In this section we prove this property, and in sections 7 and 8 we will study some of its applications. We begin with a lemma, which will be our first application of the extended Euclidean algorithm.

Lemma. *Let a, b, and c be positive integers, and assume that a and b are co-prime.*

(1) *If b divides the product ac, then b divides c.*
(2) *If a and b divide c, then the product ab divides c.*

Let's prove (1) first. We have, by hypothesis, that a and b are co-prime; that is, $\gcd(a, b) = 1$. It follows by the extended Euclidean algorithm that there exist integers α and β such that

$$\alpha a + \beta b = 1.$$

Now we come to the "abracadabra" of this proof: Multiply both sides of the equation by c. This gives

(6.1) $$\alpha ac + \beta cb = c.$$

The second term on the left-hand side is clearly divisible by b, but so is the first term. Indeed, it is divisible by ac, which is a multiple of b, by hypothesis. Thus the left-hand side of (6.1) is itself a multiple of b. Since it is equal to c, we have proved (1).

Now we use (1) to prove (2). If a divides c, then there exists an integer t such that $c = at$. But b also divides c. Since a and b are co-prime, it follows from (1) and $c = at$ that b must divide t. Thus $t = bk$ for some integer k. Hence

$$c = at = a(bk) = (ab)k$$

is divisible by ab, which is the conclusion of (2).

The lemma will be used very often, beginning with a proof of a property of the prime numbers that is found in Euclid's *Elements* as proposition 30 of Book VII. This is a very important property, so it is convenient to give it a name. We will call it the *fundamental property of prime numbers*.

Fundamental property of prime numbers. *Let p be a prime number and a and b be positive integers. If p divides the product ab, then p divides a or p divides b.*

We use the lemma to prove the property. By hypothesis, p divides ab. If p divides a, the proof is complete. Suppose that p does not divide a. But p is prime, so its only factors are 1 and p. Thus $\gcd(a, p) = 1$. Applying the lemma, we conclude that, since p divides ab, and p and a are co-prime, then p divides b.

7. The Greeks and the irrational

In this section we consider an application of the fundamental property of the prime numbers, proved in section 6. We show that if p is prime, then $\sqrt{p}$ is an irrational number. This is the first of many proofs that we will do using the method known as *proof by contradiction*.

The idea behind the method is very simple, and we often use it in everyday life. Here is a rather simple-minded example. You need a computer file that you know is on one of two disks, one of which is blue, the other red. Unfortunately, you don't remember which one, and neither of them has a label. What do you do? You put one of the disks in the computer—the blue disk, say—and look at its files. If the file you want is not there, then it is on the red disk. A more roundabout way to explain what you did is to say that you assumed your file was on the blue disk. Upon finding that it was not there, you realized that your assumption was wrong, and that the file was really on the red disk.

The reason why we expect such a strategy to work is we know that a certain fact cannot be true and false at the same time. Thus, if a file is on one of the two disks, and if it is not on the blue disk, then it must be on the red disk. Of course, in the world of everyday life, things are rarely, if ever, that clear-cut. Thus, you might be completely mistaken about the file being on one of the two disks, or worse, you might have deleted it from the disk and not even realized you'd done so. Luckily, in mathematics, things are often not so messy.

Let us see how this strategy can be applied in order to prove that $\sqrt{p}$ is irrational. But first, what does "irrational" mean in this context? Sometimes one hears that an irrational number is one that cannot be understood. But *not rational* here does not mean "impossible to understand"; it means "not a ratio". According to the *Oxford English Dictionary*, a *ratio* is

> a quantitative relation between two similar magnitudes determined
> by the number of times one contains the other.

This is almost the same as Euclid's definition 3 of Book V of the *Elements*. Unfortunately, this is the kind of definition that will not help you identify a ratio, unless you already know what a ratio is. In this respect it is much like Euclid's famous definition of a point as "that which has no part". Luckily all we need to know is that an irrational number is a real number that is *not a fraction*. Thus, we have a question to which we can quite easily apply the method of *proof by contradiction*. Do we want to prove that $\sqrt{p}$ *is not* a fraction? All we need to do is assume that it is a fraction, and try to deduce a contradiction from that. If we succeed, our assumption was wrong, and we have proved that $\sqrt{p}$ is irrational.

We must take care in setting up the proof. Recall that we will be assuming (in the hope of arriving at a contradiction) that $\sqrt{p}$ is a fraction. In other words, we are supposing that there exist integers a and b such that

$$(7.1) \qquad \sqrt{p} = \frac{a}{b}.$$

Moreover, we can assume that the fraction is in reduced form; that is, $\gcd(a, b) = 1$. Every fraction can be written in this form: Just cancel the greatest common divisor of the numerator and the denominator. It is important to assume that a/b is reduced, because this will make it easier to spot the expected contradiction.

In order to work with integers, let us square both sides of equation (7.1). We get

$$(7.2) \qquad p = a^2/b^2, \quad \text{that is,} \quad b^2 \cdot p = a^2.$$

Hence p divides a^2. By the fundamental property of prime numbers this implies that p divides a. Hence there exists an integer c such that $a = pc$. Replacing a by pc in (7.2), we have

$$b^2 \cdot p = p^2 \cdot c^2.$$

Canceling p from both sides, we see that p must divide b^2. Using the fundamental property of prime numbers again, we conclude that p divides b. Thus, we have seen that p divides a, and that p divides b. But this cannot happen, since $\gcd(a, b) = 1$. Hence we have the expected contradiction, which means that $\sqrt{p}$ is not a fraction. Therefore, $\sqrt{p}$ is an irrational number.

The existence of irrational numbers is a problem with a long and colorful history. According to the Greek historian Herodotus, geometry originated in Egypt, where the pharaoh distributed land to the people in rectangular plots on which he levied an annual tax. If the Nile swept away part of the plot, the surveyors had to be called in to calculate how much land had been lost. The owner of the plot was then eligible to a reduction of tax proportional to the area that had been lost.

For the Egyptians, interested only in practical measurements of area and similar calculations, all numbers were implicitly assumed to be fractions. It was the development of the more theoretical aspects of geometry in Ancient Greece that brought the irrational numbers to the fore.

The discovery of the irrationals is believed to have happened in the school of philosophy (or sect) founded by Pythagoras. The Pythagoreans were very interested in the development of geometry because they believed that the numbers (by which they meant integers and fractions) were the essence of the universe. Thus, one can imagine how horrified they must have felt when they realized that there were ratios of magnitudes that did not correspond to any fraction. It is said that Hypasus of Metapontum was expelled from the sect for making public this secret. The Pythagoreans, however, felt that this wasn't quite enough, so they built him a tomb, because for them he was already dead!

Inevitably, the discovery of irrationals soon became common knowledge among the philosophers. Plato, in his dialogue *Theaetetus*, says that Theodorus of Cirene had proved that the numbers $\sqrt{3}, \ldots, \sqrt{17}$ were irrational. Unfortunately, he does not say anything about this purported proof.

The proof given above for the irrationality of $\sqrt{p}$ was known to the Greeks. In chapter 23 of Book I of his *Prior Analytics*, Aristotle says that

> the diagonal of the square is incommensurate with the side, because odd numbers are equal to even if it is supposed to be commensurate.

This is a very condensed form of the proof that $\sqrt{2}$ is irrational. A more detailed proof is found in Euclid's *Elements*, proposition 117 of Book X.

8. Uniqueness of factorization

It is time to give a proof that the factorization of an integer (in the form made explicit in the theorem of section 1) is indeed unique. It will be a proof by contradiction, and it will use the fundamental property of prime numbers.

Suppose, *by contradiction*, that there exist positive integers greater than 2 that have more than one factorization, in the form of the theorem of section 1. Let n be the *smallest* positive integer with two, or more, different factorizations. Thus,

$$(8.1) \qquad n = p_1^{e_1} \ldots p_k^{e_k} = q_1^{r_1} \ldots q_s^{r_s},$$

where $p_1 < \cdots < p_k$ and $q_1 < \cdots < q_s$ are primes, and $e_1, \ldots, e_k, r_1, \ldots, r_s$ are positive integers. Moreover, we are assuming that these two factorizations are different. Note that this could happen for two different reasons. First, there could be primes in one of the factorizations that were not present in the other one. Second, even if the primes were the same in the two factorizations, the multiplicities could be different. Luckily, it does not matter which of these two possibilities actually occurs in (8.1).

We conclude, by inspection of the factorization on the left-hand side, that p_1 divides n. But $n = q_1^{r_1} \ldots q_s^{r_s}$. Repeated application of the *fundamental property of prime numbers* tells us that p_1 must divide one of the factors of $q_1^{r_1} \ldots q_s^{r_s}$. Ultimately this means that p_1 must divide one of the qs. But a prime can only divide another prime if they are equal. Therefore, p_1 has to be equal to one of the qs; say, $p_1 = q_j$, where $1 \leq j \leq s$.

Thus we can replace q_j by p_1 in the factorization on the right-hand side of (8.1):

$$n = p_1^{e_1} \ldots p_k^{e_k} = q_1^{r_1} \ldots q_j^{r_j} \ldots q_s^{r_s}$$
$$= q_1^{r_1} \ldots p_1^{r_j} \ldots q_s^{r_s}.$$

Now p_1 can be canceled, because it appears as a prime factor with positive multiplicity in both factorizations. Doing this, we obtain

$$p_1^{e_1-1} \ldots p_k^{e_k} = q_1^{r_1} \ldots p_1^{r_j-1} \ldots q_s^{r_s},$$

which are two factorizations of a positive integer, which we will call m. But these factorizations cannot be different. Indeed, n was assumed to be the *smallest* positive integer with two distinct factorizations, but $m = n/p_1 < n$. If the factorizations are equal, then we have, first of all, $j = 1$; so $p_1 = q_1$, and also $k = s$. Furthermore,

$$p_2 = q_2, \quad p_3 = q_3, \quad \ldots \quad \text{and} \quad p_k = q_k$$

and each prime must have the same multiplicity, so that

$$e_1 - 1 = r_1 - 1, \quad e_2 = r_2, \quad \ldots \quad \text{and} \quad e_k = r_k.$$

But these equalities imply that the factorizations in (8.1) are equal, which is a contradiction. Thus, the factorization in the form of the theorem of section 1 is indeed unique.

Having gone through all this trouble to prove the uniqueness of the factorization, we ought to face the fact that most people cannot even imagine a factorization that is not unique. So, once again, this seems to be one of those instances of mathematicians proving something that to everyone else is patently obvious.

The truth is quite otherwise. The only reason why we think the uniqueness of the factorization of integers is obvious is that this is the factorization we learn about first, and very early in our lives. So all our intuition about factorizations is built up entirely from this one—and it is unique. It is a little like saying that Euclidean geometry is obviously the only correct one. This is simply not so, and in these times of relativity theory and black holes, no educated person would make such a statement.

If you look at the history of mathematics in the last hundred years, you will realize that it is chock-full of examples of "number systems" whose elements admit a factorization into irreducible elements. Except that usually, this factorization is not unique. The most notorious example is related to *Fermat's Last Theorem*. This refers to the statement made by Fermat that if three integers x, y, and z satisfy

$$x^n + y^n = z^n$$

with $n \geq 3$, then $xyz = 0$. Fermat made a note in the margin of his copy of Diophantus' *Arithmetic* that said he had a marvelous proof of this fact, but the margin wasn't large enough to contain it.

The most obvious strategy, in trying to prove this result, is to factor the equation $z^n - y^n$ completely. In order to do this we must introduce complex numbers; thus

$$z^n - y^n = (z - y)(z - \zeta y) \cdots (z - \zeta^{n-1} y),$$

where $\zeta = \cos(2\pi/n) + i \sin(2\pi/n)$. It turns out that the set of complex numbers we end up with behaves very much like the integers. Every element of this set can be factored as a product of powers of irreducible elements; that is, elements that cannot themselves be factored. However, for most values of n the factorization is *not unique*. This turns out to be the main obstacle for a simple proof along these lines.

It has been suggested that Fermat's "proof" of the theorem could have been the faulty one hinted at above. In this case Fermat would have been trapped into believing that the factorization in the set of complex numbers he was using was unique, a fact we know to be false. It would not be surprising if Fermat fell victim to this error. As we noted in section 1, it was only with Gauss that the unique factorization theorem for the integers was spelled out in the explicit form we use today. Even after Gauss's *Disquisitiones*, E. Kummer proposed a proof like the one suggested above, not realizing that there was a problem, until the mistake was pointed out to him by a fellow mathematician. Not letting himself be defeated, Kummer went on to develop a method that bypassed the lack of uniqueness in the factorization. This allowed him to prove Fermat's Last Theorem for many more primes than had been possible before.

Fermat's Last Theorem was finally proved in 1995 by A. Wiles. He followed an approach that had been developed only in the preceding 10 years, and that made use of the theory of elliptic curves, about which he is an expert. For the history of the theorem prior to that, see Edwards 1977. For a good elementary introduction to the ideas behind Wiles's proof, see Gouvêa 1994.

9. Exercises

1. Are there positive integers x, y, and z such that $2^x \cdot 3^4 \cdot 26^y = 39^z$?

2. Let $k > 1$ be an integer. Show that

$$k! + 2, k! + 3, \ldots, k! + k$$

are all composite. Use this to prove that, no matter how large m is, there are always m consecutive composite integers.

3. Use Fermat's algorithm to find factors for the following integers: 175,557; 455,621; and 731,021.

4. Which of the claims below are true, and which are false?

    (1) $\sqrt{6}$ is irrational.

    (2) The sum of an irrational number and a fraction is always irrational.

    (3) The sum of two irrational numbers is always irrational.

    (4) The number $\sqrt{2} + \sqrt{3}$ is rational.

5. Show that if n is composite, then

$$R(n) = \frac{10^n - 1}{9} = \underbrace{111\ldots 11}_{n \text{ times}}$$

is also composite. These numbers are called *rep-units*.

Hint: If k is a factor of n, then $R(k)$ is a factor of $R(n)$.

6. Let $n > 0$ be a composite integer and let p be its *smallest* prime factor. Find all possible values of n for which

    (1) $p \geq \sqrt{n}$; and

    (2) $p - 4$ divides $\gcd(6n + 7, 3n + 2)$.

7. Let a and b be positive integers; their *least common multiple* is the smallest positive integer that is a multiple of both a and b. Now let

$$a = p_1^{e_1} p_2^{e_2} \ldots p_k^{e_k} \quad \text{and} \quad b = p_1^{r_1} p_2^{r_2} \ldots p_k^{r_k},$$

where $p_1 < p_2 < \cdots < p_k$ are prime numbers and the exponents $e_1, \ldots, e_k$ and $r_1, \ldots, r_k$ are greater than or equal to zero. Note that we are *not* assuming that the same primes come up in both factorizations; for example, if p_1 divides a, but not b, then $r_1 = 0$. Show that the only primes in the factorizations of $\gcd(a, b)$ and $\mathrm{lcm}(a, b)$ are $p_1, \ldots, p_k$, and find their multiplicities in each of these factorizations.

8. A positive integer n is a *perfect number* if the sum of all its factors (including 1 and n) is $2n$. For example, 6 and 28 are perfect numbers. Suppose that s is a positive integer for which $2^{s+1} - 1$ is a prime number.

    (1) Show that the factors of $2^s(2^{s+1} - 1)$ form two geometric progressions whose ratio is 2, the first beginning with 1, the second with $2^{s+1} - 1$.

    (2) Compute the sum of these factors and show that $2^s(2^{s+1} - 1)$ is a perfect number.

The above result is proposition 36 of Book IX of Euclid's *Elements*. These perfect numbers are sometimes called *Euclidean*.

The purpose of exercises 9 and 10 is to show that all *even* perfect numbers are Euclidean, that is, of the form $2^s(2^{s+1} - 1)$, where $2^{s+1} - 1$ is a prime number. This was proved by L. Euler, but the paper was published only in 1849, long after his death. The proof described below can be found in Dickson 1952, Chapter 1. It is interesting to note that all known perfect numbers are even, hence of the type already known to Euclid. It has been shown that if an *odd* perfect number exists, then it must be bigger than 10^{300}, and it must have at least eight prime factors.

9. Let n be a positive integer, and let $S(n)$ be the sum of all the factors of n, including 1 and n.

    (1) Show that r is a prime number if and only if $S(r) = r + 1$.

    (2) Show that n is a perfect number if and only if $S(n) = 2n$.

    (3) Let b_1 and b_2 be two co-prime positive integers. Show that d is a factor of $b_1 b_2$ if and only if $d = d_1 d_2$, where $d_1 = \gcd(d, b_1)$ and $d_2 = \gcd(d, b_2)$.

    (4) Use (3) to show that if b_1 and b_2 are *co-prime*, then $S(b_1 b_2) = S(b_1)S(b_2)$.

10. If n is an even perfect number, then it can be written in the form $n = 2^s t$, where $s \geq 1$ and t is odd.

    (1) Replace n by $2^s t$ in the formula $S(n) = 2n$ and use exercise 9(4) to show that 2^{s+1} must divide $S(t)$.

    (2) From (1) we have $S(t) = 2^{s+1} q$ for some positive integer q. Show that $t = (2^{s+1} - 1)q$.

    (3) We want to prove, by contradiction, that $q = 1$. Suppose that $q > 1$. It follows from (2) that t has at least three different factors, namely 1, q, and t. Hence $S(t) \geq 1 + q + t$. Show that $S(t) = 2^{s+1} q = t + q$, and find the expected contradiction.

    (4) From (3) we have $q = 1$. Inserting this in the previous formulae, we obtain $t = 2^{s+1} - 1$ and $S(t) = 2^{s+1}$. Thus $S(t) = t + 1$, and it follows from exercise 9(1) that t is a prime number.

Now bring all this together to show that $n = 2^s(2^{s+1} - 1)$, where the second factor is prime.

11. Let n be a positive integer. Denote by $d(n)$ the number of positive divisors of n. A number n is called *highly composite* if $d(m) < d(n)$ for all $m < n$. Write a program that, having a positive integer r as input, finds all highly composite numbers smaller than r. Use your program to list all highly composite numbers smaller than 5000. What do you deduce about the prime factors of these numbers from an inspection of the factorizations of the numbers in your list? Highly composite numbers were introduced and studied by the famous Indian mathematician Srinivasa Ramanujan; see Ramanujan 1927, p. 78.

12. Write a program that implements Fermat's factorization algorithm. The program should take as input any positive integer smaller than 2^{32}, and should output two of its factors or a message stating that the number is prime. Remember that Fermat's algorithm does not work properly if the input is even, so you must check that first. This is the first exercise of a sequence that ends with exercise 8 of Chapter 11.

3

Prime numbers

In the first two chapters we studied a few basic properties of the integers—without which we couldn't prove much, and two fundamental algorithms—without which we couldn't compute much. The subject of this chapter, however, is more visibly concerned with our ultimate goal: the RSA cryptosystem. Indeed, to make sure that our implementation of the RSA is secure we must be able to choose large prime numbers: two for each user. This is the problem that we begin to address in this chapter. First we consider primes obtained from polynomial, exponential, and primorial formulae. An important consequence of our study of the primorial formula will be a proof that there exist infinitely many prime numbers. The chapter ends with a discussion of the *sieve of Erathostenes*, which is the oldest known method for finding primes, and also the grandparent of all modern sieves.

1. Polynomial formulae

Most people's idea of what a "formula for primes" should be is encoded in the following definition. A function $f : \mathbb{N} \to \mathbb{N}$ is a *formula for primes* if $f(m)$ is a prime number for every $m \in \mathbb{Z}$. As we will see in this chapter, this is really too ambitious. Instead of "formulae for primes" we will search for formulae that often give primes. Since the simplest possible formulae are of the polynomial type, we might begin by asking, Are there polynomial formulae for primes?

It follows from the definition above that a polynomial

$$f(x) = a_n x^n + a_{n-1} x^{n-1} + \cdots + a_1 x + a_0,$$

with integer coefficients $a_n, a_{n-1}, \ldots, a_1, a_0$, gives rise to a *formula for primes* if $f(m)$ is prime, for every positive integer m. Let's experiment with the polynomial $f(x) = x^2 + 1$. We begin by computing $f(x)$ for a few positive integral values of x. The results are recorded in the table on page 50.

Note that if x is odd, then $f(x)$ is even. Thus $f(x)$ is always even and composite for odd values of x, except when $x = 1$, because then $f(1) = 2$. Hence, if $x > 1$ and $f(x)$ is prime, then x must be even. However, if $f(x)$ were prime for every even x, then the polynomial $f(2x)$ would be a formula for primes. Unfortunately, this is not true either; for example, $f(8) = 65$, which is a composite number. Thus, the polynomial $f(x) = x^2 + 1$ is not a formula for primes in the sense defined above. Of course this is only one example, so we

x	$f(x)$	Prime?
1	2	yes
2	5	yes
3	10	no
4	17	yes
5	26	no
6	37	yes
7	50	no
8	65	no
9	82	no
10	101	yes

may just have been unlucky in our choice of polynomial. Unfortunately, as the next result shows, this is not the case.

Theorem. *Given a polynomial $f(x)$ with integer coefficients, there are infinitely many positive integers m for which $f(m)$ is composite.*

We will prove the theorem only for polynomials of degree 2. The general case is dealt with similarly, except that the formulae are more complicated, and in the effort to understand them one might easily lose sight of the key ideas.

Let $f(x) = ax^2 + bx + c$ be a polynomial whose coefficients a, b, and c are integers. We may assume that $a > 0$. This means that $f(x)$ is always positive for large enough values of x. If $f(x)$ is composite for every positive integer x, then there is nothing to prove. Note that this can actually happen, for example if $f(x) = 4x$. Thus we may suppose that there exists a positive integer m such that $f(m)$ is prime.

Let h be any positive integer. We will compute $f(m + hp)$. A reasonable question is, Where did $m + hp$ come from? The best answer to this question seems to be the calculation below. We wish to find

$$f(m + hp) = a(m + hp)^2 + b(m + hp) + c.$$

Expanding the square and collecting the terms that contain p, we get

$$f(m + hp) = (am^2 + bm + c) + p(2amh + aph^2 + bh).$$

Note that the expression within the first bracket is equal to $f(m)$. But $f(m) = p$, so that

(1.1) $$f(m + hp) = p(1 + 2amh + aph^2 + bh).$$

Looking at (1.1), we might be tempted to conclude that $f(m + hp)$ is composite; after all, it is equal to p times an integer. That, of course, would mark the end of the proof. Unfortunately there is a gap in this argument. For $f(m + hp)$ to be composite, the expression in brackets on the right-hand side of (1.1) cannot be equal to 1. Thus we must find the values of h for which

$$1 + 2amh + aph^2 + bh > 1.$$

But this inequality is equivalent to

$$2amh + aph^2 + bh > 0.$$

Since h is positive *by hypothesis*, this last inequality holds only if

$$2am + aph + b > 0 \quad \text{that is, if} \quad h > \frac{-b - 2am}{ap}.$$

Note that $-b - 2am$ can be a positive number. This will happen if b is negative and smaller than $-2am$.

What have we proved? We showed that if $f(x) = ax^2 + bx + c$ is a polynomial with integer coefficients (and $a > 0$), and if $f(m) = p$ is prime, then $f(m + hp)$ is composite whenever $h > (-b - 2am)/ap$. In particular, there exist infinitely many positive integers x such that $f(x)$ is composite.

As we have already said, a similar proof works for polynomials of any given degree. Of course, the computation of $f(m + ph)$ is not so neat, but the major complication concerns the lower bound for h. Since we had a quadratic polynomial, the bound was obtained by solving a linear inequality, which is easily done. In general, if we begin with a polynomial of degree n, the lower bound on h will come from an inequality involving a polynomial of degree $n - 1$. This is illustrated in exercise 1, where the case of a cubic polynomial is considered. If the polynomial is of degree greater than 3, there will be no simple formula for the lower bound of h. Thus, in this case, we will be content to show that such a bound exists, even if we cannot write down an explicit formula for it. This requires a little elementary calculus; the details can be found in Ribenboim 1990, Chapter 3, section II.

The theorem means that the question we began with has a negative answer. However, we have considered only polynomials in one variable. Surprisingly, there exist polynomials in several variables, all of whose positive values are prime. The hitch is that these are polynomials in many indeterminates, so that using them to find primes is not very practical. For some examples see Ribenboim 1990, Chapter 3, section III.

2. Exponential formulae: Mersenne numbers

There are two exponential formulae of great historical importance. Both were studied by mathematicians of the seventeenth and eighteenth centuries, especially Fermat and Euler. The formulae are

$$M(n) = 2^n - 1 \quad \text{and} \quad F(n) = 2^{2^n} + 1,$$

where n is a non-negative integer. Numbers of the first form are called *Mersenne numbers*, and those of the second are called *Fermat numbers*.

The question of determining when a Mersenne number is prime goes back to the Ancient Greek mathematicians. In Pythagorean mysticism a number was called *perfect* if it was equal to half the sum of its positive factors. For example, the factors of 6 are 1, 2, 3, and 6. Adding them up we get

$$1 + 2 + 3 + 6 = 12 = 2 \cdot 6.$$

Hence 6 is a perfect number. Of course, prime numbers are never perfect. Indeed if p is prime, then its divisors are 1 and p, and $1 + p < 2p$, because $p > 1$.

Euclid knew that $2^{n-1}(2^n - 1)$ is perfect when $2^n - 1$ is prime. It is not difficult to show that all *even* perfect numbers are of this form, but this was only proved by Euler in the eighteenth century. The proofs of these results can be found in exercises 8, 9, and 10 of Chapter 2. Euclid's formula reduces the problem of finding even perfect numbers to that of finding prime Mersenne numbers.

Related as it is to obscure Pythagorean mysticism, the problem of finding perfect numbers may seem utterly irrelevant to someone living in the late twentieth century. However, the fact remains that this problem has been around for 2500 years, and it still has not been satisfactorily solved. For example, it is not known whether a perfect number must be even, but to this day no one has ever found an odd one. Of course, the task of solving so old a problem is an irresistible challenge to those who love numbers. Moreover, the fact that the problem is so difficult could mean that it is related to deep properties of the integers. This would make it even more important from the mathematician's point of view.

As we mentioned in the introduction, Marin Mersenne was a priest and amateur mathematician of the seventeenth century. The numbers of the form $2^n - 1$ owe their name to Mersenne's famous claim that they were prime when

$$n = 2, 3, 5, 7, 13, 17, 19, 31, 67, 127, \text{ and } 257;$$

and composite for all the other 44 positive primes smaller than 257.

The first thing to note is that Mersenne considered only prime exponents. Indeed, if n is composite, so is $M(n)$; supposing that $n = rs$, where $1 < r$, $s < n$, then

$$M(n) = 2^n - 1 = 2^{rs} - 1 = (2^r - 1)(2^{r(s-1)} + 2^{r(s-2)} + \cdots + 2^r + 1).$$

Hence, if r divides n, then $M(r)$ divides $M(n)$. The second thing to note is that the converse is false. In other words, if n is prime, then $M(n)$ need *not* be prime. We see from Mersenne's list that $M(11)$ should be composite, and one easily checks that

$$M(11) = 2047 = 23 \cdot 89.$$

As often happened at that time, Mersenne did not provide a proof of his statement. In 1732, Euler claimed that $M(41)$ and $M(47)$ were primes. These numbers are not in Mersenne's list, but in this case it was Euler who was wrong! The first mistake in the list was found by Pervusin and Seelhof in 1886. They discovered that $M(61)$ is prime, though it is not on the list. Other mistakes were found in later years. We now know that besides $M(61)$, the list misses the primes $M(89)$ and $M(107)$, and includes the composite numbers $M(67)$ and $M(257)$.

When Fermat wanted to show that a Mersenne number was prime, he searched for factors using a method that we will describe in Chapter 9, section 1. Nowadays we use the far more efficient *Lucas-Lehmer test*, which is studied in Chapter 9, section 4. Using this test, it was shown in January 1998 that the Mersenne number $M(3, 021, 377)$ is prime. It has 1,819,050 digits, and it is the largest known prime at the time of this writing.

3. Exponential formulae: Fermat numbers

The history of Fermat numbers is very similar to that of Mersenne numbers. Fermat knew that if $2^m + 1$ is prime, then m must be a power of 2. Thus, if one is interested in finding primes, one need only look at numbers of the form $2^{2^n} + 1$. In a letter of 1640 addressed to the Chevalier Frenicle, another amateur mathematician, Fermat computed these numbers for $n = 0, 1, \ldots, 6$; they are

$$3; 5; 17; 257; 65, 537; 4, 294, 967, 297 \text{ and } 18, 446, 744, 073, 709, 551, 617.$$

He then conjectured that all numbers of the form $2^{2^n} + 1$ are prime. Oddly enough, Fermat does not seem to have attempted to factor these numbers by a method similar to the one he used on Mersenne numbers. If he had, he would have discovered that $F(5)$ is composite. That's essentially what Euler did a hundred years later. We will study Euler's method in Chapter 9, section 2.

It is also interesting that Frenicle did not spot Fermat's mistake. After all, he too had been busy trying to factor Mersenne numbers. He wasn't by any means a mathematician of Fermat's clout, but the tone of their correspondence suggests that he would have been very pleased to find a mistake in Fermat's work. Remarkably, he seems to have agreed with Fermat on the likely truth of this conjecture.

Unlike Mersenne numbers, which have proved to be a rich source of large primes, very few Fermat primes are known. Indeed, the only known primes among Fermat numbers are $F(0), \ldots, F(4)$, which were known to Fermat himself. Of course, it is very hard to calculate Fermat numbers for "large" values of n. After all, the formula that describes these numbers is doubly exponential; that is, an exponent of an exponent.

In the previous two sections we have explored a little of the history of the most famous numbers described by exponential formulae. The proofs of the results we have mentioned will have to wait until Chapter 9. For the moment we may rejoice in the knowledge that Mersenne numbers are a good source of very large prime numbers.

However, we should point out that Fermat's method for finding factors of Mersenne numbers is very easy to explain, and not that much harder to prove. An elementary proof that Fermat himself would find congenial requires only a few clever identities; see Bressoud 1989, Chapter 3. Notwithstanding this, we will postpone the study of Fermat's method until Chapter 9. By then we will be in possession of the basic notions and theorems of group theory, which allows us to give a shorter and more transparent proof of Fermat's method. There is also the added bonus that the same ideas can then be used in a number of other applications, one of which is Euler's method for finding factors of Fermat numbers.

One of the fundamental principles of progress in mathematics is that important special problems are often solved only after the development of general methods and abstract theories that bring to the fore similarities among results that were previously thought to have little in common. These similarities, in turn, often point to unexpected applications of the new methods. That's why we generalize, to understand better and thus reach further.

4. The primorial formula

The factorial of an integer $n > 0$ is the product of all the positive numbers less than or equal to n. Similarly, we define the *primorial* $p^\sharp$ of a prime $p > 0$ to be the product of all primes smaller than or equal to p. For example, $2^\sharp = 2$ and $5^\sharp = 2 \cdot 3 \cdot 5 = 30$. Note that if p is the prime that comes after q, then

$$p^\sharp = q^\sharp p.$$

We wish to consider numbers of the form $p^\sharp + 1$. To understand why, take a look at the table below.

p	$p^\sharp$	$p^\sharp + 1$
2	2	3
3	6	7
5	30	31
7	210	211
11	2310	2311

All the numbers in the third column of the table are prime! Can this be merely a coincidence? If the question is meant to convey the hope that all numbers of the form $p^\sharp + 1$ are prime, then you ought to know that we didn't stop at 11 for nothing. Indeed,

$$13^\sharp + 1 = 30,031 = 59 \cdot 509$$

is composite.

However, although $p^\sharp + 1$ is not always prime, we can show that it does not have any factors smaller than or equal to p. We use the method of proof by contradiction. Suppose that $p^\sharp + 1$ has a prime factor $q \le p$. Since $p^\sharp$ is the product of all the positive primes up to p, it follows that q also divides $p^\sharp$. Thus q divides

$$(p^\sharp + 1) - p^\sharp = 1.$$

Hence $q = 1$, which contradicts the fact that it is prime. We conclude that the smallest factor of $p^\sharp + 1$ has to be bigger than p.

This might suggest the following algorithm for finding large primes. Suppose we know all primes up to p. Compute $p^\sharp + 1$. If it is prime, we are done. If not, find its smallest prime factor; this will have to be larger than p. Either way, we have found a prime larger than p. This is a bad approach for several reasons. The most obvious of these is the need to factor $p^\sharp + 1$. Even for rather small values of p the primorial $p^\sharp$ will be a huge number.

On the other hand, if we are lucky and $p^\sharp + 1$ is itself prime, then the problem is more approachable. A prime number of this form is called a *primorial prime*. Of course, the naive approach to testing primality proceeds by systematically trying to find a proper factor of a number. As we saw in Chapter 2, section 3, this algorithm is very inefficient. An algorithm for testing primality that is very convenient for numbers of the form $p^\sharp + 1$ will be studied in Chapter 10. In spite of this, only 16 primorial primes have been found, the largest of which corresponds to $p = 24,027$, and has 10,387 digits. Thus, the primorial formula is not a very efficient way to find large primes; luckily, that's not its only use.

5. Infinity of primes

The real reason why we have gone into so much detail about the primorial formula is that it allows us to give a very quick proof of the following fundamental result.

Theorem. *There are infinitely many prime numbers.*

The proof we describe here can be found in Euclid's *Elements* as proposition 20 of Book IX. We proceed by contradiction. Suppose that there are only finitely many prime numbers. This means that there exists a biggest prime; let's call it p. In other words, we are assuming that all numbers bigger than p are composite. However, as we saw in the previous section, the number $p^\sharp + 1$ cannot have prime factors smaller than or equal to p. Taken together, these last two statements imply that $p^\sharp + 1$ has no prime factors. But that contradicts the unique factorization theorem. Thus, there must be infinitely many prime numbers.

Many other proofs of the infinity of primes have been found. Euler's proof of 1737 is a very special case. It was the seed from which many later developments sprang, so we will sketch it here. Like Euclid's proof, this is also a proof by contradiction. So assume that there are only finitely many primes, and let p be the biggest of them all. Now consider the product

$$P = \left(\frac{1}{1 - 1/2} \right) \left(\frac{1}{1 - 1/3} \right) \left(\frac{1}{1 - 1/5} \right) \cdots \left(\frac{1}{1 - 1/p} \right),$$

where there is one term for each prime number. Of course, this product is equal to some positive real number. Moreover, by carefully multiplying the terms of the product, we can show that

(5.1) $$P = 1 + \frac{1}{2} + \frac{1}{3} + \frac{1}{4} + \frac{1}{5} + \frac{1}{6} + \dots,$$

where we now have a summand for each positive integer. This follows from the unique factorization theorem, but we will omit the proof of this equality since it depends on the multiplication of infinite series; see Hardy 1963, section 202. Though there are infinitely many summands, their sum could still be a finite number; for example, the infinite sum $1 + 1/2 + 1/2^2 + 1/2^3 + 1/2^4 + \dots$ is

B. Riemann (1826–1866).

equal to 2. But it is not difficult to see that the sum corresponding to P cannot be equal to any real number. First note that

$$\frac{1}{3} + \frac{1}{4} \geq 2 \cdot \frac{1}{4} = \frac{1}{2}$$

$$\frac{1}{5} + \frac{1}{6} + \frac{1}{7} + \frac{1}{8} \geq 4 \cdot \frac{1}{8} = \frac{1}{2}$$

$$\cdots$$

$$\frac{1}{2^{n-1}+1} + \cdots + \frac{1}{2^n} \geq 2^{n-1}\frac{1}{2^n} = \frac{1}{2}.$$

Therefore,

$$P > \frac{1}{2} + \frac{1}{3} + \frac{1}{4} + \frac{1}{5} + \frac{1}{6} + \cdots + \frac{1}{2^n} \geq n \cdot \frac{1}{2} = \frac{n}{2}$$

for any given integer $n > 0$. Thus P is bigger than any given number, so it cannot be a real number. This contradiction shows that there must be infinitely many prime numbers. For more details see Ingham 1932, theorem 1, p. 10, or Hardy and Wright 1994, Chapter XXII, section 22.1.

Of course, if there were only finitely many primes, life would be much simpler—and the world a less interesting place. The fact that primes are infinite in number poses many interesting problems. For example, what is their distribution like? As we move toward bigger and bigger numbers, does the "density" of primes increase or decrease? Is there a way to measure this "density"? The best way to state this problem is by using the π function. Let x be a positive real number, and denote by $\pi(x)$ the quantity of positive primes less than or equal to x. An important problem in number theory is finding good estimates for $\pi(x)$.

Mention the distribution of primes to a mathematician, and you'll soon hear the name Riemann. Building on the ideas unleashed by Euler's proof of the infinity of primes, B. Riemann wrote what was to become the seminal work on $\pi(x)$ and related questions. This paper, published in 1859, contains many interesting and fascinating results, many of which were stated without proof. Unfortunately, Riemann died of consumption seven years later, never having had the time to work out the details of his proofs. This task was undertaken by several mathematicians, notably J. Hadamard.

One of the fruits of Hadamard's effort to fill up the gaps left by Riemann was a proof of the famous *prime number theorem*. This theorem says that

$$\lim_{x \to \infty} \frac{\pi(x) \log x}{x} = 1,$$

where $\log x$ is the logarithm of x to base e. This result is even older than Riemann's paper; it had originally been conjectured by Gauss. It was proved in 1896, independently by J. Hadamard and C. J. de la Vallé-Poussin.

Loosely speaking, the prime number theorem says that if x is very large, then $\pi(x)$ is approximately equal to $x/\log(x)$. But the approximation will only be good if x is truly enormous. For example, if $x = 10^{16}$, then

$$\pi(x) - \left[\frac{x}{\log x} \right] = 7{,}804{,}289{,}844{,}393$$

which is of the order of magnitude of 10^{13}. Since in this case $x/\log x$ is of the order of 10^{14}, we have quite a large error. There are many other simple functions that give good approximations for $\pi(x)$ when x is large. One of these is studied experimentally in exercise 11. For a detailed discussion of the distribution of prime numbers see Hardy and Wright 1994, Chapter XXII, and Ingham 1932. For the history of the prime number theorem see Bateman and Diamon 1996.

6. The sieve of Erathostenes

The *sieve of Erathostenes* is the oldest known method for finding primes. Unlike the methods discussed in the previous sections, it does not use any special formula. Erathostenes was a Greek mathematician born around 284 B.C. He was very proficient in many branches of knowledge, but his contemporaries believed that he hadn't reached a truly eminent position in any of them. So they nicknamed him "Beta" (the second letter of the Greek alphabet) and "Pentatlos". That a mathematician whose work has survived for 2300 years could have been known by these names is a good measure of the greatness of the mathematics of Ancient Greece.

In his *Arithmetic*, published around A.D. 100, Nicomachus of Gerasa introduces the sieve of Erathostenes as follows:

> The method for obtaining these [the prime numbers] is called
> by Erathostenes a sieve, since we take the odd numbers mixed
> together and indiscriminate, and out of them by this method, as
> though by some instrument or sieve, we separate the prime and
> indecomposable by themselves, and the secondary and composite
> by themselves.

For a longer quotation from Nicomachus on the sieve see Thomas 1991, p. 101. Thus the sieve has this name because, when it is applied to a list of positive integers, the composite numbers pass through, but the primes are retained. Let's see how it works.

First of all, the purpose of the sieve is to determine all positive prime numbers smaller than an upper bound $n > 0$, which we assume to be an integer. To perform the sieve with pencil and paper we proceed as follows. First, write a list of all *odd* integers between 3 and n. The reason we leave the even numbers out is that the only even prime is 2.

Now we begin to sieve the list. The first number in the list is 3. Beginning at the next number in the list (which is 5), we cross out every third number from the list. Having done this, we will have crossed out all the multiples of 3, bigger than 3 itself, that were listed.

Now pick the smallest number in the list, bigger than 3, that hasn't been crossed out. It is 5, and the number next to it is 7. We then cross out every fifth number from the list beginning at 7. That way, all multiples of 5 will be crossed out. We carry on this procedure until we get to n and stop. Note that if we are about to cross out every pth number from the list, then we always begin counting from $p + 2$, even when this number has already been crossed out at a previous loop of the sieve.

For example, if $n = 41$, the list of odd integers is

$$
\begin{array}{cccccccccc}
3 & 5 & 7 & 9 & 11 & 13 & 15 & 17 & 19 & 21 \\
23 & 25 & 27 & 29 & 31 & 33 & 35 & 37 & 39 & 41.
\end{array}
$$

Having crossed out every third number beginning with 5, we have

$$
\begin{array}{cccccccccc}
3 & 5 & 7 & \not{9} & 11 & 13 & \not{15} & 17 & 19 & \not{21} \\
23 & 25 & \not{27} & 29 & 31 & \not{33} & 35 & 37 & \not{39} & 41.
\end{array}
$$

Now we cross out every fifth number beginning with 7, thus getting

$$
\begin{array}{cccccccccc}
3 & 5 & 7 & \not{9} & 11 & 13 & \not{15} & 17 & 19 & \not{21} \\
23 & \not{25} & \not{27} & 29 & 31 & \not{33} & \not{35} & 37 & \not{39} & 41.
\end{array}
$$

We would now have to cross out every seventh number beginning with 9. But if we do that, no further numbers will be crossed out. Next, we would have to cross out every eleventh number beginning with 13, but once again that has no effect on the list. Indeed, none of the numbers left in the list will be crossed out at any later stage of the sieving process. Thus the positive odd primes smaller than 41 are

$$
\begin{array}{cccccccccccc}
3 & 5 & 7 & 11 & 13 & 17 & 19 & 23 & 29 & 31 & 37 & 41.
\end{array}
$$

There are a couple of things to notice in this example. First, though we said that one should go on sieving up to the upper limit n (41 in the example), we had already gotten rid of all composite numbers by the time we sieved the multiples of 5. All the sieving done after that was redundant. Second, some numbers were crossed out more than once. This is the case with 15, for example. It was first crossed out when we sieved for multiples of 3. But it is also a multiple of 5, so it was crossed out again when we sieved for multiples of 5.

Let's see what can be done to improve the efficiency of the sieve in light of these two remarks. Let's deal first with the second of the above remarks; that is, can we arrange things so that every number is crossed out only once? Unfortunately, the answer is that there is no efficient way of doing so. However, we can improve matters a little.

Suppose that we are about to sieve for the multiples of a prime p. According to our previous description of the sieve, we should cross out every pth number starting with $p + 2$—the number next to p in the list. A simple improvement is to begin crossing out not from $p + 2$, but from the smallest multiple of p *that is not a multiple of a prime smaller than* p. Let's find this number. The positive multiples of p are numbers of the form kp, where k is a positive integer. If $k < p$, then kp is also a multiple of a number smaller than p, namely k. Thus, the first multiple of p that is not a multiple of a prime smaller than p is p^2. So it is enough to cross out every pth number beginning with p^2. However, we must call attention to the fact that, even after this improvement, there are numbers that are crossed out more than once.

As for the other remark, Can we stop sieving before we reach n? This time the answer is yes, and it follows from what we have just done. For example, suppose that we are about to cross out every pth number. As we have just seen, the first number we have to cross out is p^2. However, if $p^2 > n$, this number is outside the list, and we can forget about it. Thus we need only cross out every pth number so long as $p \leq \sqrt{n}$. Since p is an integer, this is equivalent to $p \leq [\sqrt{n}]$. In the example above, $[\sqrt{41}] = 6$. This explains why sieving for multiples of 3 and 5 was enough to catch all composite numbers in the list.

We must now discuss how to program the sieve in a computer. The list of odd numbers is represented by a vector (or array). Remember that there are two numbers associated with every entry of a vector. One of the numbers is the value of the entry; the other identifies the position of this entry in the vector. This last number is the index of the position. For example, in the vector

$$(\quad a \quad b \quad c \quad d \quad e \quad f \quad g \quad)$$
$$\uparrow$$

the value of the entry marked by the arrow is b, and its index is 2 since it is the second entry of the vector.

Let us go back to the sieve of Erathostenes. Suppose we wish to find all primes smaller than a positive odd integer n. We must first construct a vector with $(n - 1)/2$ entries, one for each odd integer between 3 and n. Thus the entry indexed by j corresponds to the odd integer $2j + 1$. The entries will have one of two possible values: 1 or 0. If the value of an entry is 0, then the odd number it represents has been crossed out at some previous stage of the sieving process. Thus, at the moment we start the sieve, each entry is initialized with 1, because no numbers have been crossed out yet. To "cross out" the number $2j + 1$ corresponds to replacing with a 0 the 1 that was the original value of the jth entry of the vector. Of course, this entry could have been crossed out in a previous loop of the sieve. In that case its value is already 0 and will not be changed when we perform any later loop.

We now give a more or less detailed version of the algorithm for the sieve of Erathostenes that was described above. This version includes both of the improvements previously discussed. Thus every pth number is crossed out beginning with p^2, and the algorithm stops when $p > \sqrt{n}$.

Sieve of Erathostenes

Input: an odd positive integer n
Output: the list of all odd positive primes less than or equal to n

Step 1 Begin by creating a vector $\mathbf{v}$ with $(n-1)/2$ entries, each of which will be initialized with 1, and letting $P = 3$.
Step 2 If $P^2 > n$, write the list of numbers $2j + 1$ for which the jth entry of the vector $\mathbf{v}$ is 1 and stop; otherwise go to step 3.
Step 3 If the entry indexed by $(P-1)/2$ of the vector $\mathbf{v}$ is 0, increase P by 2 and return to step 2; otherwise go to step 4.
Step 4 Give to a new variable T the value P^2; replace with 0 the value of the entry indexed by $(T-1)/2$ of the vector $\mathbf{v}$ and increase T by $2P$; repeat these two steps until $T > n$, then increase P by 2 and return to step 2.

Note that in the last step we increased T by $2P$ instead of P, which is what you may have expected. We did this because the vector $\mathbf{v}$ represents a list of odd numbers so that both T and P are odd. Thus, if we are crossing out every Pth number, then the number that will be crossed out after T is $T + 2P$.

It may have occurred to you that there is a simple change in the procedure above that will clearly speed up the algorithm. The way we have been getting rid of the unwanted composite numbers in the vector consists in marking their position with a 0 instead of a 1. However, since we don't really care for composite numbers, why don't we simply delete them from the vector? Unfortunately we cannot do that. The trouble is, the way we know that the value of an entry in the vector $\mathbf{v}$ is a multiple of p depends on its position. In other words, multiples of p occur at every pth position in the vector. If we delete some numbers from the list, that will cease to hold, and the algorithm that we have described will not work.

Like all algorithms, the sieve of Erathostenes has limitations. For example, it is not an efficient way to search for very large primes. Remember, however, that the purpose of the algorithm is to find *all primes* smaller than a certain upper bound. That is clearly not feasible if the bound is too large.

Keeping within the limits set by the algorithm's purpose, we notice two weak points: The sieve requires a lot of memory space, and it must run for too many loops. On the positive side, we needn't compute a single division, and the sieve is very easily programmed.

7. Exercises

1. Let a, b, c, and d be integers, with $a > 0$, and consider the polynomial $f(x) = ax^3 + bx^2 + cx + d$ of degree 3. Suppose that there exists a positive integer m such that $f(m) = p > 0$ is prime. Find the positive integer values of h for which $f(m + hp)$ is composite.

2. Using the trial division algorithm of Chapter 2, section 2, find all prime factors of $p^\# + 1$ for

(1) $p = 17$.
(2) $p = 13$.

An odd prime is either of the form $4n + 1$ or of the form $4n + 3$. In other words, the possible remainders of the division of an odd prime by 4 are 1 or 3. For example, 3, 7, 11, and 19 are of the form $4n + 3$, while 5 and 13 are of the form $4n + 1$. The purpose of exercises 3 through 7 is to give a proof that there are infinitely many prime numbers of the form $4n + 3$. It is also true that there are infinitely many primes of the form $4n + 1$, but the proof is not quite so elementary; it can be found in Hardy and Wright 1994, section 2.3, theorem 13.

3. Show that the product of two integers of the form $4n + 1$ is of the form $4n + 1$.

4. Show that every odd prime number is either of the form $4n + 1$ or of the form $4n + 3$.

5. Is the product of two numbers of the form $4n + 3$ also a number of the form $4n + 3$?

6. Suppose that $3 < p_1 < \cdots < p_k$ are primes of the form $4n + 3$. Using exercise 3, show that $4(p_1 \ldots p_k) + 3$ must be divisible by a prime of the form $4n + 3$ that does not belong to the set $\{3, p_1, \ldots, p_k\}$.

7. Using the previous exercise, show that there exist infinitely many primes of the form $4n + 3$.

8. We saw in Chapter 1, exercise 5, that if $n > m$ are positive integers, then $\gcd(F(n), F(m)) = 1$. In other words, two distinct Fermat numbers cannot have a common factor. Use this fact to give another proof that there are infinitely many primes.

9. Show that if p, $p + 2$, and $p + 4$ are all positive primes, then $p = 3$.

10. Let f be a quadratic polynomial. Write a program to find the integer values of n, smaller than 100, for which $f(n)$ is prime. The input of the program will be the coefficients a, b, and c of the polynomial $f(x) = ax^2 + bx + c$. These coefficients are integers and can be positive or negative. The program will compute $f(n)$ for all non-negative integers n smaller than 100 and find which of these are prime. To do this we must first run the sieve of Erathostenes to find all the primes smaller than $\max\{|f(0)|, |f(100)|\}$. Note that it will be necessary to impose limits on the size of $|a|$, $|b|$, and $|c|$, otherwise $f(x)$ can fall outside the range of integers supported by the computer language you are using. Apply the program to each of the following polynomials:

(1) $f(x) = x^2 + 1$.
(2) $f(x) = x^2 - 69x + 1231$.
(3) $f(x) = 2x^2 - 199$.
(4) $f(x) = 8x^2 - 530x + 7681$.

The second polynomial is a variant of a famous example published by L. Euler in 1772.

11. We mentioned in section 5 that there are several formulae that give an approximation for $\pi(x)$, the number of primes smaller than or equal to x. For example, as a consequence of the prime number theorem, we have that $x/\log x$ is approximately equal to $\pi(x)$ when x is large. But in this case, x has to be truly enormous for the error to be small. In this exercise we study experimentally another formula that gives a better approximation when x is small. The formula is

$$S(x) = \frac{x}{\log x}\left(1 + \left[\sum_{k=0}^{12} a_k (\log\log x)^k\right]^{-1/4}\right),$$

where $\log$ denotes the logarithm to the base e, and

$a_0 = 229,168.50747390, \quad a_1 = -429,449.7206839, \quad a_2 = 199,330.41355048,$

$a_3 = 28,226.22049280 \quad a_4 = 0, a_5 = 0, \quad a_6 = -34,712.81875914,$

$a_7 = 0, \quad a_8 = 33,820.10886195, \quad a_9 = -25,379.82656589,$

$a_{10} = 8,386.14942934, \quad a_{11} = -1,360.44512548, \quad a_{12} = 89.14545378.$

Use the sieve of Erathostenes as the base for a program that, taking an integer $x > 0$ as input, computes $\pi(x)$. Use this program to compute $\pi(x) - S(x)$ for $x = 11; 100; 1000; 2000; 3000; \ldots; 9000;$ and $10,000$. Compare with the corresponding values of $\pi(x) - x/\log x$. What do you conclude?

12. We have seen that an odd prime is either of the form $4n + 1$ or of the form $4n + 3$. Moreover, it follows from exercise 7 that there are infinitely many primes of the form $4n + 3$. It is also true that there are infinitely many primes of the form $4n + 1$, though the proof is more difficult—see the comments before exercise 3. The purpose of this exercise is to study experimentally the relative frequency of these two types of primes. Let x be a positive real number. Let $\pi_1(x)$ be the number of positive primes of the form $4n + 1$ smaller than or equal to x, and let $\pi_3(x)$ be the corresponding number of primes of the form $4n + 3$. Write a program, based on the sieve of Erathostenes, to compute $\pi_1(x)$ and $\pi_3(x)$ when the input x is a positive integer. Use the program to compute $\pi_1(x)$, $\pi_3(x)$ and $\pi_1(x)/\pi_3(x)$ for $x = 100k$ and $1 \le k \le 10^5$. It is known that $\lim_{x\to\infty} \pi_1(x)/\pi_3(x) = 1$. Do your data support this result?

13. Adapt the program of exercise 12 to determine the smallest positive integer x for which $\pi_1(x) > \pi_3(x)$.

The numerical data available at the beginning of the century led some mathematicians to the conclusion that, except for small values of x, the inequality $\pi_1(x) < \pi_3(x)$ would always hold. The truth emerged in 1914, when J. E. Littlewood showed that there are infinite sequences $x_1, x_2, \ldots$ and $y_1, y_2, \ldots$ of positive real numbers such that

$$\lim_{i\to\infty} (\pi_1(x_i) - \pi_3(x_i)) = \infty \quad \text{and} \quad \lim_{i\to\infty} (\pi_1(y_i) - \pi_3(y_i)) = -\infty.$$

The moral is clear: It is dangerous to generalize from numerical data.

4

Modular arithmetic

Most of the algorithms of the previous chapters prove divisibility by performing the division and checking that the remainder is zero. Now it has been shown, by a method that we will study in Chapter 9, that $5 \cdot 2^{23,473} + 1$ is a factor of $F(23,471)$. Since these numbers are very large, we expect that even checking the truth of the statement will take a very long time. So far, so good; but how many digits does $F(23,471)$ have? Using logarithms, it is easy to show that it has more than 10^{7063} digits! Thus, $F(23,471)$ has more digits than there are particles in the visible universe. Needless to say, we will not be able to show that $5 \cdot 2^{23,473} + 1$ is a factor of $F(23,471)$ by performing the division. How is it done, then?

The way out of this dilemma is to use *modular arithmetic*, which is the theme of this chapter. This is the basic technique for dealing with questions of divisibility, but it is also useful for calculations related to periodic phenomena, as we will see in Chapter 7.

Although the basic ideas of modular arithmetic had been around for some time, they were first developed systematically by Gauss at the very beginning of his *Disquisitiones arithmeticæ*; see Gauss 1986. Nowadays the subject is usually approached from the point of view of *equivalence relations*, a topic we will review in detail in section 1.

1. Equivalence relations

The best way to introduce modular arithmetic is by using equivalence relations. Since these relations will play an important role here and elsewhere in the book, it is a good idea to go through the basic concepts in some detail.

Suppose that X is a set, which may be finite or infinite. A *relation* in X is a rule that specifies how the elements of this set are to be compared. This is not a formal definition, but it is quite satisfactory for our purposes. Note that in order to define a relation we must say what the subjacent set is; in other words, we have to be clear about which elements are to be compared.

Let's consider a few examples. In the set of integers, there are many simple relations, like *equality*, *inequality*, *less than*, and *less than or equal to*. In a set of colored balls we have the relation *same color*. The latter is a very good example to keep in mind because it is very concrete. By the way, we are assuming that each ball in the set has been painted one color only—multicolored balls are not allowed.

Equivalence relations are relations of a very special kind. Going back to the general setup, suppose that X is a set in which a relation has been defined. It is convenient to have a symbol to denote this relation; let's call it $\sim$. Now $\sim$ is an *equivalence relation* if, for every $x, y, z \in X$, the following properties hold:

(1) $x \sim x$.

(2) If $x \sim y$, then $y \sim x$.

(3) If $x \sim y$ and $y \sim z$, then $x \sim z$.

The first property is called the *reflexive* property. It says that when we have an equivalence relation, we can always compare an element with itself. This holds for the equality of integers: Every integer is equal to itself. But it does *not* hold for the relation $<$. Hence $<$ in $\mathbb{Z}$ is *not* an equivalence relation.

The second property is called the *symmetric* property. The relation $<$ in the set of integers is not symmetric. Indeed, $2 < 3$, but it is not true that $3 < 2$. Note that the relation $\leq$ in the set $\mathbb{Z}$ is reflexive, but *not* symmetric.

The third is the *transitive* property. In the set of integers, the relations "equality", "less than", and "less than or equal to" are all transitive. But the inequality of integers is not transitive. Indeed, $2 \neq 3$ and $3 \neq 2$ does not imply that $2 \neq 2$. Note that $\neq$ is symmetric, but not reflexive.

We have been careful to give examples of relations for which these properties are false because this is the only way to understand what the properties really mean. It is familiarity with examples (both pro and con) that makes us comfortable in handling concepts. There is no dearth of examples of equivalence relations. The equality of integers clearly satisfies all three properties above, so it is an equivalence relation. The relation "same color" in a set of colored balls is also another simple and very concrete example. Examples in a set of polygons include the relations "same number of sides" and "same area".

Equivalence relations are used to classify elements of a given set that have similar properties by grouping them into subsets. The natural subdivisions of a set produced by an equivalence relation are called *equivalence classes*. Thus, let X be a set, and suppose that $\sim$ is an *equivalence relation* defined in X. Let x be an element of X. The *equivalence class* of x is the subset of all elements of X that are equivalent to x under $\sim$. Denoting the equivalence class of x by $\overline{x}$, we have

$$\overline{x} = \{y \in X : y \sim x\}.$$

Here is a simple example. Let $\mathcal{B}$ be a set of colored balls with the equivalence relation "same color". The equivalence class of a red ball in $\mathcal{B}$ is the set of *all* red balls contained in $\mathcal{B}$.

There is a property of the equivalence classes that is so important that we will call it the *basic principle* of equivalence classes. It says that *any element of an equivalence class is a good representative of the whole class*. In other words, if you know one element of an equivalence class, you can immediately reconstruct the whole class. This is obvious if we consider the set $\mathcal{B}$ of colored balls with the relation "same color". Suppose you are told that a paper bag contains the elements of an equivalence class of $\mathcal{B}$. You ask to see one element of the set; it is a blue ball, say. You immediately conclude that the bag contains the equivalence class of all blue balls of $\mathcal{B}$. It couldn't be easier!

Let's go back to the set X with the equivalence relation $\sim$. The basic principle says that if y is an element of the equivalence class of x, then the equivalence classes of x and y are the same. In other words,

$$\text{if } x \in X \text{ and } y \in \overline{x}, \text{ then } \overline{x} = \overline{y}.$$

Let's prove this directly from the properties used to define an equivalence relation. If $y \in \overline{x}$, then, by the definition of an equivalence class, we must have $y \sim x$. The symmetric property implies that $x \sim y$. But if $z \in \overline{x}$, then we also have $z \sim x$. Hence, by the transitive property, $z \sim y$. Thus $z \in \overline{y}$. We have shown that $\overline{x} \subseteq \overline{y}$. A similar argument proves that $\overline{y} \subseteq \overline{x}$. It may all seem a little pedantic, perhaps. But it is such a source of confusion and puzzlement that we should spare no trouble to make it clear what the principle really means—and that it is a straightforward consequence of the definition of an equivalence relation. Speaking of being pedantic, have you realized that the reflexive property is behind the fact that $x \in \overline{x}$?

The basic principle is related to an important property of equivalence relations. As before, let X be a set with an equivalence relation $\sim$; then

(1) X is the union of its equivalence classes with respect to $\sim$; and

(2) two equivalence classes that are different cannot have a common element.

The first property follows from the fact, already mentioned, that the equivalence class of an element x contains x itself. To prove the second property, suppose that $x, y, z \in X$ and that $z \in \overline{x} \cap \overline{y}$. Since $z \in \overline{x}$, it follows that $\overline{z} = \overline{x}$ by the basic principle. Similarly, $\overline{z} = \overline{y}$. So $\overline{x} = \overline{y}$. Note that (1) and (2) allow us to break the set X into disjoint subsets, the equivalence classes. This is called a *partition* of X.

The set formed by the equivalence classes of X with respect to the equivalence relation $\sim$ has a special name: the *quotient set* of X by $\sim$. Note that the elements of the quotient set are subsets of X. Therefore the quotient set is *not* a subset of X. This can be a source of much puzzlement, so beware.

Let's end this section with an example in which the true nature of fractions is at last revealed. What are fractions made of? When you look at a fraction, what you see are two integers, one of which (the denominator) must be non-zero. Of course, you probably think of that as a quotient. But if you are pressed, you may try to choose the easy way out and say that a fraction is indeed a pair of integers, the second of which is non-zero. However, that cannot be correct.

In mathematics, two pairs are equal if they have the same first elements and the same second elements. So the pairs $(2, 4)$ and $(1, 2)$ are *not* equal. But the fractions $2/4$ and $1/2$ are equal; so fractions are not pairs of integers after all.

What are fractions, then? They are elements of a quotient set. Consider the set $\mathcal{Q}$ of pairs of integers (a, b), with $b \neq 0$. In the usual jargon, $\mathcal{Q} = \mathbb{Z} \times (\mathbb{Z} \setminus \{0\})$. Two pairs, (a, b) and (a', b'), are now said to be equivalent if $ab' = a'b$. One easily checks that this is an equivalence relation. A fraction is an equivalence class of $\mathcal{Q}$ with respect to this relation. Hence a/b stands not for the pair (a, b), but for the infinite set of all pairs in $\mathcal{Q}$ equivalent to (a, b). Thus the set $\mathbb{Q}$ of rational numbers is the quotient set of $\mathcal{Q}$ by the equivalence relation just defined.

Imagine for a moment that you have never heard of fractions before. So all you have to go on is the description above. If you are now told that you will have to calculate with fractions, you may feel you have good reason to panic. After all, you have just learned that a fraction is an infinite set. The thought of adding an infinite set to another infinite set does seem rather worrisome. This is the point where the basic principle of equivalence classes comes to the rescue. You needn't carry the burden of the whole infinite set; all you need to know is an element of this set. This element tells you all you need to know about the whole equivalence class. Moreover, any element of the class will do.

Thus, you may calculate with $1/2$ as you always did, just as if it were a pair of integers. You are only reminded that a fraction is an equivalence class of pairs when, in the midst of a calculation, you realize that you can *simplify* a fraction. At that moment you are replacing a representative of the equivalence class with another that will make your calculation easier to handle.

Why this long digression about fractions? In the next section we will define an equivalence relation in the set $\mathbb{Z}$, and the quotient set of this relation will play an absolutely fundamental role in this book. Just as with fractions, the equivalence classes will be infinite sets—and we will have to calculate with them. But now you know that there is no cause for worry.

2. The congruence relation

Let's analyze the familiar 24-hour clock in the framework of the previous section. First, when someone says "one o'clock" we cannot tell whether the person means that time today, yesterday, or tomorrow. Thus "one o'clock" is not a moment in time; it is an equivalence class of such moments. Let's be more explicit. First divide the time continuum into equal intervals, called *hours*. Then define an equivalence relation: Two moments that differ by 24 of these equal intervals are equivalent. Now one o'clock is an equivalence class of moments for this special equivalence relation. This may sound like complicating the obvious, but it can be quite useful if you are dealing with cyclic phenomena.

We will now consider a similar equivalence relation defined in the set of integers. Choose a positive integer n that will be fixed from now on. It is called the period or *modulus* of the relation that we are about to define.

Now we construct an equivalence relation in the set of integers. We do this by declaring that every nth integer (beginning with 0) is equivalent under this relation. In other words, any two integers that differ by a multiple of n are equivalent. More formally, two integers a and b are *congruent modulo* n if $a - b$ is a multiple of n; in that case we write

$$a \equiv b \pmod{n}.$$

Here are a few numerical examples. If $n = 5$ is the modulus, then

$$10 \equiv 0 \pmod 5 \quad \text{and} \quad 14 \equiv 24 \pmod 5.$$

Let's choose a different modulus, say $n = 7$; in this case

$$10 \equiv 3 \pmod 7 \quad \text{and} \quad 14 \equiv 0 \pmod 7.$$

Note that numbers that are congruent for a certain modulus need not be congruent for a different modulus. Thus 21 is congruent to 1 modulo 5; but these numbers are *not* congruent modulo 7 because $21 - 1 = 20$ is not a multiple of 7.

We must now check that congruence modulo n is an equivalence relation. First, the reflexive property: To check it we must show that if a is any integer, then $a \equiv a \pmod n$. This will be true if $a - a$ is a multiple of n. But $a - a = 0$ is a multiple of any integer. So the congruence modulo n is reflexive.

Next comes the symmetric property. Suppose that $a \equiv b \pmod n$ for some integers a and b. This means that $a - b = kn$ for some integer k. Multiplying this equation by -1, we obtain

$$b - a = -(a - b) = (-k)n,$$

which is also a multiple of n. Thus $b \equiv a \pmod n$, and we have proved that the congruence modulo n is symmetric.

Finally, the transitive property: Suppose that $a \equiv b \pmod n$ and $b \equiv c \pmod n$, where a, b, and c are integers. By definition, these congruences say that $a - b$ and $b - c$ are multiples of n. But if we add up multiples of n, we obtain a multiple of n. Thus $(a - b) + (b - c) = (a - c)$ is a multiple of n. In other words, $a \equiv c \pmod n$, as we wished to show. Having verified that these three properties hold for the congruence modulo n, we conclude that it is an equivalence relation.

The set that will occupy most of our attention throughout the rest of this chapter is the quotient set of $\mathbb{Z}$ by the congruence modulo n. It is called the *set of integers modulo* n, and it is denoted by $\mathbb{Z}_n$. By the definition of the quotient set we know that the elements of $\mathbb{Z}_n$ are subsets of $\mathbb{Z}$, that is, equivalence classes of $\mathbb{Z}$ for the congruence modulo n. We wish to identify these classes.

Let $a \in \mathbb{Z}$. The class of a is formed by the integers b such that $b - a$ is a multiple of n. In other words, $b - a = kn$, for some $k \in \mathbb{Z}$. Thus we may describe the equivalence class of a by

$$\bar{a} = \{a + kn : k \in \mathbb{Z}\}.$$

Note that $\bar{0}$ is the set of all multiples of n, and that every one of these equivalence classes is an infinite set.

We have seen that $\bar{a}$ has infinitely many elements, all of which are good representatives of the whole class. Thus it is reasonable to ask, Is there a simple way to find the smallest positive integer that represents $\bar{a}$? The answer is yes; all one has to do is divide a by n. Let r be the remainder and q the quotient of this division; then

$$a = nq + r \quad \text{and} \quad 0 \le r \le n - 1.$$

Hence $a - r = nq$ is a multiple of n. Therefore $a \equiv r \pmod{n}$. The number r is called the *residue* of a modulo n.

We have actually proved more than we bargained for. Indeed, we have shown that any given integer is congruent to an integer between 0 and $n - 1$. In particular, the quotient set $\mathbb{Z}_n$ has at most the n equivalence classes $\overline{0}, \ldots, \overline{n-1}$. To make sure that there are exactly n *distinct* equivalence classes modulo n we must show that no two of these can be equal. But each congruence class can be represented by a non-negative integer smaller than n. If they were equal, their representatives would have to be congruent modulo n. In other words, the difference of two non-negative integers smaller than n would have to be a multiple of n. That cannot happen, so the classes $\overline{0}, \ldots, \overline{n-1}$ are indeed distinct. Hence

$$\mathbb{Z}_n = \{\overline{0}, \overline{1}, \ldots \overline{n-1}\}.$$

An equivalence class $\overline{a}$ is said to be written in *reduced form* if $0 \leq a \leq n-1$. As in the case of fractions, it is always convenient to represent a congruence class in reduced form. There are two reasons for this. First, it is obviously easier to choose a smaller representative for a class than a bigger one. Second, if two classes are in reduced form, then it is very easy to decide whether they are equal or not: They will be equal if and only if their representatives are equal. Of course this is *not* true if the classes are not in reduced form.

This is all very well, but it seems quite abstract. It would be good to have a nice geometric picture, a way of drawing the set $\mathbb{Z}_n$. Let's start by recalling the picture we have of the set $\mathbb{Z}$. Most people think of the integers as a sequence of points equally spaced along an infinite straight line. Somewhere along the line a point represents 0, so that we see the negative integers to the left of it and the positive integers to the right. Now, the congruence modulo n identifies every nth integer, so that when we get to n we should actually be back at 0.

If you think of the integer line as being flexible, you could perhaps pick the point marked n and glue it at 0. That would give a circle. If you keep on wrapping the line around this circle, you'll see that numbers that are congruent modulo n will occupy the same points on the circumference of the circle. So $\mathbb{Z}_n$ is to be pictured as a circumference around which the n congruence classes are marked at regular intervals.

3. Modular arithmetic

The geometric picture described at the end of section 2 helps us give a simple description of addition in $\mathbb{Z}_n$. Think of the n equivalence classes of $\mathbb{Z}_n$ as if they were the hours marked on the face of a clock. Let's assume that $\overline{0}$ is the point uppermost in the circle (the 12 o'clock mark), while the other classes are disposed clockwise around the circle at regular intervals. This clock will have one hand only, fixed at the center of the circle, so that we can point it to the classes.

We want to make this "clock" into a machine for calculating sums in $\mathbb{Z}_n$. Arguably the "clock" is to modular arithmetic as finger-reckoning is to "normal" arithmetic. The latter, by the way, has a long and distinguished history. It was taught at the convent schools in the Middle Ages, and one of the first English books on arithmetic, Robert Recorde's *The Grounde of Artes*, had a whole section explaining "the arte of Numbrynge by the hande". So we are in very good company.

So, suppose we want to add $\bar{a}$ and $\bar{b}$, two classes of $\mathbb{Z}_n$. We will assume that both classes are in reduced form, so a and b are non-negative and smaller than n. To find $\bar{a} + \bar{b}$ proceed as follows. Place the hand of the "clock" at the point marked $\bar{a}$, then move it b places in the clockwise direction. The hand will now be pointing to the result of the sum $\bar{a} + \bar{b}$.

Here's a numerical example. Suppose we want to add $\bar{5}$ to $\bar{4}$ in $\mathbb{Z}_8$. First place the hand of the "clock" at $\bar{5}$. Now move it clockwise four places. When you do that, the hand goes past $\bar{0}$ and stops at $\bar{1}$. Hence $\bar{4} + \bar{5} = \bar{1}$ in $\mathbb{Z}_8$.

The problem with our machine is that it is impracticable if n is big. Thus, a more mathematical way of calculating sums in $\mathbb{Z}_n$ is required. It is quite simple, actually. Let $\bar{a}$ and $\bar{b}$ be the classes of $\mathbb{Z}_n$ that we want to add up. Then the operation is defined by the formula

$$\bar{a} + \bar{b} = \overline{a + b}.$$

Some care is needed in interpreting this formula. On the left-hand side we have the sum of two classes of $\mathbb{Z}_n$; on the right-hand side we have the class that corresponds to the sum of two integers. Thus the addition of classes is defined in terms of an operation that we already know, the addition of integers.

Let's go back to the example of addition in $\mathbb{Z}_8$ that we calculated using the machine. We wish to add $\bar{5}$ to $\bar{4}$. According to the formula, we add the integers 4 and 5 first; since their sum is 9, it follows that $\bar{5} + \bar{4} = \bar{9}$. This seems to give a different result from the one obtained with the machine. But don't forget that $9 - 1 = 8$, so that $\bar{9} = \bar{1}$.

This last example points to an important problem. We have seen that an equivalence class can be represented by any one of its elements; this is the basic principle of section 1. But to add two classes we first add their representatives, and then take the corresponding class. How can we be sure that if different representatives are chosen, the resulting classes will still be the same? Just to make sure that you got the point, consider one more time the sum of $\bar{5}$ and $\bar{4}$ in $\mathbb{Z}_8$. Following the rule set up before, we found that this sum was equal to $\bar{9}$. However, $\overline{13} = \bar{5}$ and $\overline{12} = \bar{4}$. But the formula says that if we add $\overline{13}$ to $\overline{12}$, we get $\overline{25}$. Thus it may seem at first that, by choosing different representatives for the classes, we get a different sum. But this is only apparently so, because in fact $25 - 9 = 16$ is a multiple of 8. So $\overline{25} = \bar{9}$ in $\mathbb{Z}_8$.

One way to solve the problem would be to write the classes in reduced form before we add them up. This is impractical and wholly unnecessary. As the example above suggests, the result of the sum is always the same no matter which representatives are chosen for the classes. This is very important and must be checked in detail. Let $\bar{a}$ and $\bar{b}$ be two classes in $\mathbb{Z}_n$. Suppose that $\bar{a} = \bar{a'}$ and $\bar{b} = \bar{b'}$. We wish to show that $\overline{a + b} = \overline{a' + b'}$. But $\bar{a} = \bar{a'}$ means that $a - a'$ is a multiple of n, and the same holds for $b - b'$. The sum of multiples of n is a multiple of n, so that

$$(a - a') + (b - b') = (a + b) - (a' + b')$$

must be a multiple of n. Hence, $\overline{a + b} = \overline{a' + b'}$, as we wished to prove.

The subtraction of classes can de defined in an analogous way and offers no difficulty. Let's see how multiplication is to be defined. Suppose that $\overline{a}$ and $\overline{b}$ are classes in $\mathbb{Z}_n$. The definition of addition of classes suggests that we should have

$$\overline{a} \cdot \overline{b} = \overline{ab}.$$

As with addition, we must make sure that this definition does not depend on the choice of representatives for the classes. Thus, suppose that $\overline{a} = \overline{a'}$ and $\overline{b} = \overline{b'}$. We must show that $\overline{ab} = \overline{a'b'}$. Since $\overline{a} = \overline{a'}$, it follows that $a - a'$ is a multiple of n; say, $a = a' + rn$ for some integer r. Similarly, $b = b' + sn$ for some integer s. Multiplying a by b we get

$$ab = (a' + rn)(b' + sn) = a'b' + (a's + rb' + srn)n.$$

Hence $ab - a'b'$ is a multiple of n, so that $\overline{ab} = \overline{a'b'}$.

Now that we know how to add and multiply classes, we must consider whether these operations behave as their namesakes in $\mathbb{Z}$. Let $\overline{a}$, $\overline{b}$, and $\overline{c}$ be classes in $\mathbb{Z}_n$. The addition of classes satisfies the following properties:

$$(\overline{a} + \overline{b}) + \overline{c} = \overline{a} + (\overline{b} + \overline{c})$$

$$\overline{a} + \overline{b} = \overline{b} + \overline{a}$$

$$\overline{a} + \overline{0} = \overline{a}$$

$$\overline{a} + \overline{-a} = \overline{0}$$

The element $\overline{-a}$ is called the *symmetric* of $\overline{a}$. Note that if we assume that $\overline{a}$ is in reduced form, then the reduced form of $\overline{-a}$ is $\overline{n - a}$. The multiplication of classes satisfies the following properties:

$$(\overline{a} \cdot \overline{b}) \cdot \overline{c} = \overline{a} \cdot (\overline{b} \cdot \overline{c})$$

$$\overline{a} \cdot \overline{b} = \overline{b} \cdot \overline{a}$$

$$\overline{a} \cdot \overline{1} = \overline{a}$$

Thus every property of multiplication corresponds to a property of addition, with one exception. The exception is the existence of a symmetric element. We will come back to this question in section 7, where we will discuss the division of classes.

There is also the *distributive* property

$$\overline{a} \cdot (\overline{b} + \overline{c}) = \overline{a} \cdot \overline{b} + \overline{a} \cdot \overline{c}.$$

These properties follow quite easily from their counterparts for the addition and multiplication of integers, so we will omit the proofs. The conscientious reader should have no difficulty in working them out.

So far, so good. The operations in $\mathbb{Z}_n$ apparently behave just like their namesakes in $\mathbb{Z}$. That, however, is quite dangerous, because it may give one a false sense of security. Indeed, one key property of the integers does not hold for congruences modulo n. An example will make this clear. Consider the classes $\overline{2}$ and $\overline{3}$ in $\mathbb{Z}_6$. They are both different from $\overline{0}$. However,

$$\overline{2} \cdot \overline{3} = \overline{6} = \overline{0}.$$

Thus the product of non-zero elements of $\mathbb{Z}_6$ can be zero. This, of course, cannot happen in $\mathbb{Z}$.

An important consequence of this example is that in $\mathbb{Z}_n$ one cannot always cancel a non-zero element. In other words, it is *not* always true that if $\overline{a} \neq 0$, then

$$\overline{a} \cdot \overline{b} = \overline{a} \cdot \overline{c} \quad \text{implies that} \quad \overline{b} = \overline{c}.$$

Thus, for example, $\overline{2} \cdot \overline{3} = \overline{2} \cdot \overline{0}$, but $\overline{3} \neq \overline{0}$. We will return to these questions in section 7, but first we will consider some applications of modular arithmetic.

4. Divisibility criteria

Most people remember from elementary school that a number is divisible by 3 if the sum of its digits is divisible by 3. But why is this true? We can prove this easily using congruences modulo 3. Recall that a number is divisible by 3 if and only if it is congruent to 0 modulo 3. Thus, in the language of modular arithmetic, the divisibility criterion for 3 says that a number is congruent to 0 modulo 3 if and only if the same is true of the sum of its digits. It is this last statement that we now prove.

Let a be an integer, and let $a_0, a_1 \ldots, a_n$ be its decimal digits. In other words,

$$a = a_n 10^n + a_{n-1} 10^{n-1} + \cdots + a_1 10 + a_0,$$

where $0 \leq a_i \leq 9$ for $i = 0, \ldots, n$. We were careful to show, in the previous section, that multiplication is independent of the representatives of the classes modulo n. Thus, since $10 \equiv 1 \pmod 3$, it follows that 10^k is congruent to 1^k modulo 3 for any positive integer k. In other words, any power of 10 has residue 1 modulo 3. Therefore,

$$a \equiv a_n + a_{n-1} + \cdots + a_1 + a_0 \pmod 3,$$

from which it immediately follows that $a \equiv 0 \pmod 3$ if and only if $a_n + a_{n-1} + \cdots + a_1 + a_0 \equiv 0 \pmod 3$. But this is just what we needed to prove.

Note that all the calculations above hold if we replace 3 with 9, because $10 \equiv 1 \pmod 9$. Therefore, an integer is divisible by 9 if and only if the sum of its digits is divisible by 9. Let's apply a similar argument to other numbers, for example, 11.

Once again, we will assume that $a = a_n 10^n + a_{n-1} 10^{n-1} + \cdots + a_1 10 + a_0$, where $a_n, a_{n-1}, \ldots, a_1, a_0$ are the digits of a. Note that $10 \equiv -1 \pmod{11}$, so that

$$10^k \equiv (-1)^k \pmod{11}$$

is either 1 (if k is even) or -1 (if k is odd). Thus

$$a \equiv a_n(-1)^n + a_{n-1}(-1)^{n-1} + \cdots + a_2 - a_1 + a_0 \pmod{11}.$$

In plain English the criterion says that a number is divisible by 11 if and only if the alternating sum of its digits is divisible by 11. For example, 3443 is divisible by 11 because $3 - 4 + 4 - 3 = 0$ is divisible by 11.

The divisibility criteria for 2 and 5 are too obvious to require further justification. Thus we have found simple divisibility criteria for all prime numbers between 2 and 11, except 7. Let's find out what happens if we apply the same approach to 7.

We already know, from the previous examples, that the part of the argument that depends on the modulo is the computation of the powers of 10. Now $10 \equiv 3$ (mod 7), and powers of 3 are not quite so simple to compute as powers of 1 or -1. Let's try some examples. All congruences below are taken to be modulo 7.

$$10^2 \equiv 3^2 \equiv 2$$
$$10^3 \equiv 10 \cdot 10^2 \equiv 3 \cdot 2 \equiv 6 \equiv -1$$
$$10^4 \equiv 10 \cdot 10^3 \equiv (-1) \cdot 3 \equiv 4$$
$$10^5 \equiv 10 \cdot 10^4 \equiv 3 \cdot 4 \equiv 5$$
$$10^6 \equiv 10 \cdot 10^5 \equiv 3 \cdot 5 \equiv 1$$

Note that the last residue above is equal to $10^0 = 1$. This means that the residues will get repeated in cycles of 6. The calculations above show that the divisibility criterion for 7 will be a lot more difficult to remember than the ones for 3 and 11. Since we've come this far, we may as well explicitly state the criterion in a simple case. Suppose that $a = a_2 10^2 + a_1 10 + a_0$, where $0 \le a_0, a_1, a_2 \le 9$. Using the residues for powers of 10 calculated above, we have

$$a \equiv a_2 10^2 + a_1 10 + a_0 \equiv 2a_2 + 3a_1 + a_0 \pmod{7}.$$

Thus a is divisible by 7 if and only if $2a_2 + 3a_1 + a_0$ is divisible by 7. For example, 231 is divisible by 7 because $2 \cdot 2 + 3 \cdot 3 + 1 = 14$ is divisible by 7.

5. Powers

In many applications we will be faced with the following problem. Let a, k, and n be positive integers; find the remainder of the division of a^k by n. If k is very large, it may not even be possible to compute the digits of a^k, as in the example at the beginning of the chapter. However, we can get around this problem using modular arithmetic.

Let's begin with a simple example. Suppose that we want to find the remainder of the division of 10^{135} by 7. We have seen in the previous section that $10^6 \equiv 1 \pmod 7$. Dividing 135 by 6, we find that $135 = 6 \cdot 22 + 3$. Thus we have the following congruences modulo 7:

$$10^{135} \equiv (10^6)^{22} \cdot 10^3 \equiv (1)^{22} \cdot 10^3 \equiv 6.$$

Hence the remainder of the division of 10^{135} by 7 is 6.

It is not always this easy. For example, what is the remainder of the division of 3^{64} by 31? Calculating a few powers of 3 modulo 31, we quickly find that $3^3 \equiv -4 \pmod{31}$. Instead of computing higher powers in the hope that one of them will turn out to be 1, let's make use of the information we already have. Since $64 = 3 \cdot 21 + 1$, we obtain the following congruences modulo 31:

$$3^{64} \equiv (3^3)^{21} \cdot 3 \equiv (-4)^{21} \cdot 3 \equiv -(2)^{42} \cdot 3.$$

We haven't yet obtained the remainder, but it is only a power of 2 that lies between us and our aim. Luckily, $2^5 \equiv 1 \pmod{31}$. Since $42 = 8 \cdot 5 + 2$, we find that

$$3^{64} \equiv -(2)^{42} \cdot 3 \equiv -(2^5)^8 \cdot 2^2 \cdot 3 \equiv -12 \pmod{31}.$$

But $-12 \equiv 19 \pmod{31}$, so the remainder of the division of 3^{64} by 31 is 19.

Would the calculations have turned out to be easier if we had pressed on with the computation of the powers of 3 until we got to 1? The answer is no, as you will realize if you try to find the smallest positive integer r for which $3^r \equiv 1 \pmod{31}$.

Suppose we now want to find the remainder of the division of 6^{35} by 16. In this case it is no use trying to find the smallest power of 6 that is congruent to 1 modulo 16, because this power does not exist. Indeed,

$$6^4 \equiv 2^4 \cdot 3^4 \equiv 0 \cdot 3^4 \equiv 0 \pmod{16},$$

and so

$$6^{35} \equiv 6^4 \cdot 6^{31} \equiv 0 \pmod{16}.$$

These examples illustrate some of the tricks we use to simplify the calculation of the residues of powers modulo n. More tricks will be found in the coming chapters. Of course a computer does not require any tricks. That's not to say that computers do not use modular arithmetic for such calculations; indeed they do. A very fast algorithm for calculating powers modulo n can be found in section 2 of the Appendix. This algorithm can be used to show that $5 \cdot 2^{23,473} + 1$ is a factor of $F(23,471)$, thus achieving what at first seemed an impossible task.

6. Diophantine equations

Next, we use congruences to show that certain diophantine equations do not have solutions. A *diophantine equation* is a polynomial equation in several unknowns, with integer coefficients. Examples are $3x - 2y = 1$, $x^3 + y^3 = z^3$, and $x^3 - 117y^3 = 5$. When we talk of finding solutions of diophantine equations, we usually mean integer solutions. These equations derived their name from the Greek mathematician Diophantus of Alexandria, who lived around A.D. 250. In his *Arithmetic*, Diophantus discusses in detail the problem of finding solutions to indeterminate equations. However, he looked for rational solutions rather than integer solutions, as we usually do today.

Since these are equations in several variables, they may have infinitely many integer solutions. For example, for any integer k, the numbers $x = 1 + 2k$ and $y = 1 + 3k$ satisfy the equation $3x - 2y = 1$. The equation $x^3 + y^3 = z^3$ is a particular case of Fermat's Last Theorem, referred to in the introduction and at the end of Chapter 2. As we have seen, if x, y, and z are integers that satisfy this equation, then one of them must be zero. This special case of the theorem was first proved by L. Euler in 1770. You may recall that it was in the margin of his copy of Diophantus' *Arithmetic* that Fermat first stated his Last Theorem.

The equation $x^3 - 117y^3 = 5$ has a more recent and humble history. In a paper written in 1969, D. J. Lewis showed that this equation could not have more than 18 integer solutions. Two years later, R. Finkelstein and H. London proved that the equation does not have any integer solutions after all. Their proof is short—it fills only page 111 of volume 14 of the *Canadian Mathematical Bulletin*—but it is not elementary. However, in 1973, F. Halter-Koch and V. Şt. Udresco independently gave a proof that the equation has no solution that uses only congruence modulo 9. We will describe this proof in detail.

The proof is by contradiction. Suppose that the equation $x^3 - 117y^3 = 5$ has an integer solution. This means that there exist integers x_0 and y_0 such that $x_0^3 - 117y_0^3 = 5$. Since all these numbers are integers, we may reduce this last equation modulo 9. But 117 is divisible by 9, so that

$$x_0^3 \equiv x_0^3 - 117y_0^3 \equiv 5 \pmod 9.$$

Hence, if the given equation has integer solutions x_0 and y_0, then $x_0^3 \equiv 5$ (mod 9). Is this possible? To find out, recall that every integer modulo 9 has a residue between 0 and 8. Thus, it will be enough to compute the cube modulo 9 of each one of these residues.

classes modulo 9	$\bar{0}$	$\bar{1}$	$\bar{2}$	$\bar{3}$	$\bar{4}$	$\bar{5}$	$\bar{6}$	$\bar{7}$	$\bar{8}$
cubes modulo 9	$\bar{0}$	$\bar{1}$	$\bar{8}$	$\bar{0}$	$\bar{1}$	$\bar{8}$	$\bar{0}$	$\bar{1}$	$\bar{8}$

A simple inspection of the table shows that the cube of an integer modulo 9 must have 0, 1, or 8 as its residue. In particular, there cannot be an integer x_0 such that $x_0^3 \equiv 5$ (mod 9). Therefore, $x^3 - 117y^3 = 5$ cannot have integer solutions, as we wanted to prove.

We may draw a moral from the history of this example: *The first proof of a theorem is often neither the simplest nor the most elegant.* This happens because the first proof is usually found by the explorer, who is breaking new ground. With time, the connections between the new methods and neighboring areas become clearer, and that makes it possible to find a shortcut, a proof simpler and more direct than the first one. The example we have given is a rather naive one, but of course that doesn't invalidate the conclusion we have drawn. As the mathematician A. S. Besicovitch once said, "A mathematician's reputation rests on the number of bad proofs he has given".

7. Division modulo n

It is time to return to the problem of division of classes in $\mathbb{Z}_n$. But first let us consider the same question in a more familiar setting. Let a and b be real numbers. One way to divide a by b is to multiply a by $1/b$. The number $1/b$ is called the inverse of b, and it is uniquely defined as the solution of the equation $b \cdot x = 1$. From a practical point of view, this is not really an improvement because, to find $1/b$, we still have to divide 1 by b. However, from a conceptual point of view, it is sometimes better to talk about inverses than about division. Finally, $1/b$ exists only if $b \neq 0$, because the equation $0 \cdot x = 1$ does not have a solution. With this in mind, we now turn to $\mathbb{Z}_n$.

As always, n is a fixed positive integer, and suppose that $\overline{a} \in \mathbb{Z}_n$. We say that $\overline{\alpha} \in \mathbb{Z}_n$ is the *inverse* of $\overline{a}$ if the equation $\overline{a} \cdot \overline{\alpha} = \overline{1}$ holds in $\mathbb{Z}_n$. It is clear that $\overline{0}$ does not have an inverse in $\mathbb{Z}_n$. Unfortunately, $\overline{0}$ may not be the only element of $\mathbb{Z}_n$ that does not have an inverse. We must consider this point in greater detail.

Suppose that $\overline{a} \in \mathbb{Z}_n$ has an inverse $\overline{\alpha}$, and let's see what we get from this. From the equation

$$\overline{a} \cdot \overline{\alpha} = \overline{1},$$

it follows that $a\alpha - 1$ is divisible by n. In other words,

(7.1) $$a\alpha + kn = 1$$

for some integer k. Note that it follows from (7.1) that $\gcd(a, n) = 1$. Thus we conclude that if $\overline{a}$ has an inverse in $\mathbb{Z}_n$, then $\gcd(a, n) = 1$.

Is the converse of this last statement true? To find out, suppose that a is an integer such that $\gcd(a, n) = 1$. Equation (7.1) suggests that we should apply the extended Euclidean algorithm to the numbers a and n. Doing so we find integers α and β such that

$$a\alpha + n\beta = 1.$$

But this equation is equivalent to

$$\overline{a} \cdot \overline{\alpha} = \overline{1}$$

in $\mathbb{Z}_n$. Hence the class $\overline{\alpha}$ computed with the help of the extended Euclidean algorithm is the inverse of $\overline{a}$ in $\mathbb{Z}_n$. Therefore, if $\gcd(a, n) = 1$, then $\overline{a}$ is invertible in $\mathbb{Z}_n$. We sum it up in a theorem.

Invertibility theorem. *The class $\overline{a}$ is invertible in $\mathbb{Z}_n$ if and only if the integers a and n are co-prime.*

The argument above is a constructive proof of the invertibility theorem, in the sense that it gives a procedure for checking whether the inverse exists and for finding it, if it does. Of course, this procedure is just a straightforward application of the extended Euclidean algorithm. For example, does $\overline{3}$ have an inverse in $\mathbb{Z}_{32}$? If so, what is it? Applying the extended Euclidean algorithm to

32 and 3, we find that $\gcd(3, 32) = 1$, and that

$$3 \cdot 11 - 32 = 1.$$

So the inverse exists. Rewriting the equation modulo 32, one gets $\overline{3} \cdot \overline{11} = \overline{1}$. Thus $\overline{11}$ is the inverse of $\overline{3}$ in $\mathbb{Z}_{32}$.

The set of invertible elements of $\mathbb{Z}_n$, denoted by $U(n)$, will play a key role in Chapters 8, 9, and 10. By the invertibility theorem we have

$$U(n) = \{\overline{a} \in \mathbb{Z}_n : \gcd(a, n) = 1\}.$$

It is very easy to compute $U(p)$ when p is a prime number because, in this case, the condition $\gcd(a, p) = 1$ is equivalent to p does not divide a. Since this is true of all positive integers smaller than p, it follows that $U(p) = \mathbb{Z}_p \setminus \{\overline{0}\}$.

But this happens *only if p is prime*. If n is composite, and $1 < k < n$ is one of its factors, then $\gcd(k, n) = k \neq 1$, so $\overline{k}$ is not invertible in $\mathbb{Z}_n$. Here are two simple examples:

$$U(4) = \{\overline{1}, \overline{3}\} \quad \text{and} \quad U(8) = \{\overline{1}, \overline{3}, \overline{5}, \overline{7}\}.$$

A key property of $U(n)$ is that the product of two of its elements is again an element of $U(n)$. In other words, if $\overline{a}$ and $\overline{b}$ are invertible classes of $\mathbb{Z}_n$, then $\overline{a} \cdot \overline{b}$ is also an invertible class. This will be very important in Chapter 8, so let's check it in detail. Suppose that $\overline{a}$ has inverse $\overline{\alpha}$ and $\overline{b}$ has inverse $\overline{\beta}$ in $\mathbb{Z}_n$. The inverse of $\overline{a} \cdot \overline{b}$ is $\overline{\alpha} \cdot \overline{\beta}$; indeed,

$$(\overline{a} \cdot \overline{b})(\overline{\alpha} \cdot \overline{\beta}) = (\overline{a} \cdot \overline{\alpha})(\overline{b} \cdot \overline{\beta}) = \overline{1} \cdot \overline{1} = \overline{1}.$$

We will have a lot more to say about $U(n)$ in later chapters.

Let's go back to the problem of dividing $\overline{a}$ by $\overline{b}$ in $\mathbb{Z}_n$, with which we started the section. First, we need to know whether $\overline{b}$ is an invertible element of $\mathbb{Z}_n$. If it is, we use the extended Euclidean algorithm to find its inverse. Let's call the inverse $\overline{\beta}$. To divide $\overline{a}$ by $\overline{b}$ we compute the product $\overline{a} \cdot \overline{\beta}$. For example, let's divide $\overline{2}$ by $\overline{3}$ in $\mathbb{Z}_8$. Applying the extended Euclidean algorithm to 3 and 8, we have $\gcd(3, 8) = 1$, and $\overline{3}$ is its own inverse. Thus the result of dividing $\overline{2}$ by $\overline{3}$ in $\mathbb{Z}_8$ is $\overline{6}$.

We can use the results of this section to solve linear congruences in $\mathbb{Z}_n$. A *linear congruence* is an equation of the form

(7.2) $$ax \equiv b \pmod{n},$$

where $a, b \in \mathbb{Z}$. If this were a linear equation over the reals, we would divide it through by a. We will try to use the same idea here. Assuming that $\gcd(n, a) = 1$, we conclude that, by the invertibility theorem, there exists $\alpha \in \mathbb{Z}$ such that $\alpha a \equiv 1 \pmod{n}$. Multiplying both sides of equation (7.2) by α, we have

$$x \equiv \alpha a x \equiv \alpha b \pmod{n},$$

and the equation has been solved. For example, to solve $7x \equiv 3 \pmod{15}$ we must first find the inverse of 7 modulo 15. Since $15 - 2 \cdot 7 = 1$, the inverse of 7 modulo 15 is $-2 \equiv 13 \pmod{15}$. Multiplying the equation $7x \equiv 3 \pmod{15}$ by 13, we have

$$x \equiv 13 \cdot 3 \equiv 39 \equiv 9 \pmod{15},$$

which is the solution we were looking for.

Note that the method used to solve linear congruences shows that if $\gcd(a, n) = 1$, then the congruence $ax \equiv b \pmod{n}$ has one and only one solution modulo n. In other words, though infinitely many integers are solutions of this equation, they are all congruent modulo n. This may sound trite, but it is not; indeed, it can be false if $\gcd(a, n) \neq 1$. The equation $2x \equiv 1 \pmod{8}$, for example, has no solutions at all. We will come back to this question at the beginning of Chapter 7.

8. Exercises

1. The relations below are defined in the set $\mathbb{Z}$. Which of them are equivalence relations?

(1) $a \sim b$ when $\gcd(a, b) = 1$.

(2) Fix an integer $n > 0$, and define $a \sim b$ if and only if $\gcd(a, n) = \gcd(b, n)$.

2. Find the residue of a modulo n for each of the integers a and n below:

(1) $a = 2351$ and $n = 2$.

(2) $a = 50{,}121$ and $n = 13$.

(3) $a = 321{,}671$ and $n = 14$.

3. Compute the residues of each of the following powers:

(1) $5^{20} \pmod{7}$.

(2) $7^{1001} \pmod{11}$.

(3) $81^{119} \pmod{13}$.

(4) $13^{216} \pmod{19}$.

4. Find the remainder of the division of $1000!$ by 3^{300}.

5. Compute $U(n)$ and find the inverse of each one of its elements when $n = 4$, 11, and 15.

6. Solve the following linear congruences:

(1) $4x \equiv 3 \pmod{4}$.

(2) $3x + 2 \equiv 0 \pmod{4}$.

(3) $2x - 1 \equiv 7 \pmod{15}$.

7. Show that every element of $U(34)$ is a power of $\overline{3}$.

8. Show that the Diophantine equation $x^2 - 7y^2 = 3$ does not have integer solutions.

9. Show that $p = 274{,}177 = 1071 \cdot 2^8 + 1$ is a prime factor of the Fermat number $F(6)$. Hint: First compute 1071^8 modulo p. To do this note that $1071 = 7 \cdot 9 \cdot 17$, compute the eighth power of each of these factors modulo p, and then multiply them out. Now, since $p = 1071 \cdot 2^8 + 1$, we have $(1071 \cdot 2^8)^8 \equiv 1 \pmod{p}$. On the other hand, $(1071 \cdot 2^8)^8 \equiv 1071^8 \cdot 2^{64} \pmod{p}$. Replace 1071^8 in this last formula by its residue modulo p and compare with the previous congruence. The fact that p divides $F(6)$ comes out of this as if by magic.

The next three problems are closely related. We will return to them in the context of a primality test for Mersenne numbers; see Chapter 9, section 4.

10. Consider the set of real numbers of the form $a + b\sqrt{3}$, where a and b are integers. This set is usually denoted by $\mathbb{Z}[\sqrt{3}]$. Since the elements of $\mathbb{Z}[\sqrt{3}]$ are real numbers, they can be added and multiplied. Show that if $\alpha, \beta \in \mathbb{Z}[\sqrt{3}]$, then $\alpha + \beta$ and $\alpha\beta$ both belong to $\mathbb{Z}[\sqrt{3}]$.

11. Define a relation in $\mathbb{Z}[\sqrt{3}]$ as follows. Fix an integer n, and let $\alpha, \beta \in \mathbb{Z}[\sqrt{3}]$. We say that $\alpha \equiv \beta \pmod{n}$ if and only if there exists $\gamma \in \mathbb{Z}[\sqrt{3}]$ such that $\alpha - \beta = n\gamma$. Show that this is an equivalence relation of $\mathbb{Z}[\sqrt{3}]$.

12. Fix a positive integer n, and let $\mathbb{Z}_n[\sqrt{3}]$ be the quotient set of $\mathbb{Z}[\sqrt{3}]$ by the equivalence relation of the previous exercise. Let $\widetilde{\alpha}$ and $\widetilde{\beta}$ be classes in $\mathbb{Z}_n[\sqrt{3}]$. Show that the rules

$$\widetilde{\alpha} + \widetilde{\beta} = \widetilde{\alpha + \beta}$$

$$\widetilde{\alpha}\widetilde{\beta} = \widetilde{\alpha\beta}$$

determine well-defined operations in $\mathbb{Z}_n[\sqrt{3}]$.

13. In this exercise we show that a number of the form $4n + 3$ cannot be written as a sum of two squares of integers.

    (1) Show that the square of an integer can only be congruent to 0 or 1 modulo 4.

    (2) Use (1) to show that if x and y are integers, then $x^2 + y^2$ can only be congruent to 0, 1, or 2 modulo 4.

    (3) Use (2) to show that an integer of the form $4n + 3$ cannot be written as the sum of two squares.

This result is a particular case of a theorem communicated by Fermat to Roberval in a letter of 1640. Fermat also knew that every prime of the form $4n + 1$ can be written as the sum of two squares. For more details see Weil 1987, Chapter II, section VIII. See also exercise 14 of Chapter 5.

14. Write a program that implements the algorithm for computing powers modulo n that is described in section 2 of the Appendix. The input will consist of three positive integers a, k, and n; the output will be the residue of a^k modulo n. This algorithm will be fundamental for all the applications of later chapters.

5

Induction and Fermat

Having learned the basic facts of modular arithmetic, we are now ready to return to the study of prime numbers. The main result of this chapter is a very useful theorem first proved by Fermat. This theorem is in fact a straightforward consequence of a much deeper theorem in group theory, which we will learn about in Chapter 8. In this chapter, however, we follow Fermat's lead and give a direct proof of the theorem using *finite induction*. It is with a description of this method of proof that we begin.

1. Hanoi! Hanoi!

Have you ever played with a puzzle called the *Towers of Hanoi?* It consists of three vertical pegs stuck to a wooden base, and a number of wooden discs (six in the version I own). Each disc has a hole through which a peg can be passed. Let's call the three pegs **A**, **B**, and **C**. The discs have different diameters, and when we begin to play they should all be on peg **A**, stacked in decreasing order of size. The biggest disc should be at the bottom and the smallest one at the top of a neat tower, as shown in the figure.

The problem is to shift the whole tower from peg **A** to peg **C**, using **B** as intermediary, but subject to the following constraints:

(1) Only one disc can be moved at a time; and
(2) A bigger disc can never be placed on top of a smaller one.

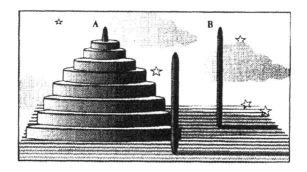

The Towers of Hanoi.

Note that (1) means that only the top disc of any stack can be moved at any given time. Of course these constraints would make the problem insoluble if we did not have the extra peg **B**.

It is a good idea to try to solve the puzzle for yourself, in order to get the hang of it. With practice one can do it fairly quickly. But the question we wish to pose goes beyond that: Can we find a formula for the minimum number of moves required to shift a tower of n discs from **A** to **C**? Of course, we are assuming that the discs will be moved according to the rules.

This problem has a very important application; that is, so long as you are prepared to believe the following tale. Under the great dome of a temple in India there are three diamond needles, each as thick as the body of a bee. At the moment of creation, God placed on one of these needles 64 discs of pure gold, the biggest one at the bottom, the other ones forming a tower with the smallest disc on top. To the priests who preside at the temple He gave the task of moving the discs according to the rules we have stated above. When the whole tower of 64 discs finally gets moved to one of the other needles, God will return and, with a bang, put an end to the world. Thus, to find out when the world will end all we have to do is solve the problem of the minimum number of moves for the 64 discs.

Setting aside the eschatological aspects of the problem, let's return to the wooden set we began with. If the set had only one disc, it would be enough to move it from **A** to **C**. We haven't broken any rules, and the puzzle is solved. So one movement is enough in this case. Now suppose that we have two discs. First we move the smaller disc to peg **B**; now the bigger disc can be moved to **C**; finally, the smaller disc is moved to **C**, so that it rests on top of the bigger disc. Thus three moves are enough to solve the puzzle with two discs. If you have a set, it may be a good idea to count the moves as you try to move a tower with four and then with five discs.

Let's now deal with the general case of a puzzle with n discs. The argument will be easier to digest if we describe it in the form of a dialogue between pupil and teacher.

Teacher: Let's assume that the discs have been numbered $1, 2, \ldots, n$, from top to bottom. So the smallest disc is number 1 (and sits on the top) and the biggest (the one on the bottom of the stack) is number n. What do we have to do in order to be able to move disc n?

Pupil: Huh?

Teacher: We want to move disc n, but we have all these discs sitting on top of it. What do we have to do?

Pupil: Remove all the ones that are on top of it?

Teacher: That's it, so we have to remove the $n - 1$ discs that lie above it. Let's not forget that we want to move all the discs to peg **C**, and disc n will have to be the one at the bottom of the stack. Where would you rather move the other $n - 1$ discs to?

Pupil: To peg **B**?

Teacher: To peg **B**. However, there is a problem.

Pupil: There always is.

Teacher: The rules. By rule (1) we can only move one disc at a time; by rule (2) the $n - 1$ discs must be stacked in decreasing order of size in peg **B**. What do we have to do to shift the $n - 1$ smaller discs from **A** to **B**?

Pupil: We have to move one disc at a time, without breaking the rules.

Teacher: More precisely?

Pupil: It would be like solving the puzzle with $n - 1$ discs, I guess. I mean, shifting the tower of $n - 1$ discs to **B**, instead of **C**.

Teacher: What about the intermediary peg?

Pupil: Might it be **C**?

Teacher: That's it. Summing up: We have to move disc n to **C**. But we can only do that if we first remove to **B** the $n - 1$ discs that lie above it. This is done by playing the $n - 1$ smaller discs using **C** as intermediary. Thus the whole stack of $n - 1$ discs gets shifted from **A** to **B**. Having done that, we are free to move disc n to **C**.

Pupil: How do we then move the $n - 1$ smaller discs to **C**, so that they come to lie on top of disc n? I mean, would we do that by playing the $n - 1$ discs again, so that we move them (one by one?) from **B** to **C**?

Teacher: It may take a long time, but that's the way it is done. Note that we've had to solve the puzzle with $n - 1$ discs *twice*. First we moved the top $n - 1$ discs from **A** to **B** (using **C** as intermediary). This leaves no disc on top of n. Then we move disc n to **C**. Next we play the $n - 1$ discs one more time, from **B** to **C** (using **A** as intermediary). Having done that, all the n discs got stacked on peg **C** and none of the rules were broken.

Pupil: Ugh!

Teacher: It is not finished!

Pupil: Isn't it? Oh, dear.

Teacher: We want to find out the minimum number of moves we have had to make, don't we?

Pupil: I guess we still do.

Teacher: To make the argument easier to follow, let's call $T(n)$ the *minimum* number of moves required to solve the puzzle with n discs. But we've already seen that to move disc n we have first to shift the $n - 1$ discs that lie on top of it. How many moves to have them out of the way?

Pupil: To get them to peg **B** we had to solve the puzzle with $n - 1$ discs, didn't we? So I guess we made at least $T(n - 1)$ moves.

Teacher: Since we moved the $n - 1$ smaller discs to peg **B**, that means that peg **C** has no discs on it. So we can now move disc n to **C**. How many moves to get there?

Pupil: One?

Teacher: I mean on the whole; how many moves since we began to play?

Pupil: Oh, $T(n - 1) + 1$?

Teacher: What next?

Pupil: We still have to move the $n - 1$ smaller discs from **B**, where they now are, to **C**, where they'll be on top of disc n.

Teacher: That's it. And how many moves to reach that goal?

Pupil: Not less than $T(n - 1)$, surely, because that's the minimum number of moves for the puzzle with $n - 1$ discs.

Teacher: So we see that, from the moment we began to play, we have made a total of $T(n - 1) + 1 + T(n - 1) = 2T(n - 1) + 1$ moves. Moreover, if we look carefully at the way we've been arguing, we realize that we couldn't possibly complete the puzzle with fewer moves. So what is $T(n)$?

Pupil: The minimum number of moves required to solve the puzzle with n discs and get the tower shifted from **A** to **C**.

Teacher: Yes, but how do we compute $T(n)$, assuming that we already know $T(n - 1)$?

Pupil: $T(n) = 2T(n - 1) + 1$?

Teacher: That's it. So we can now find the minimum number of moves required if we are to solve the puzzle with six discs.

The final outcome of the dialogue is the formula $T(n) = 2T(n - 1) + 1$. Note that this formula does *not* tell us directly what $T(n)$ is. To find out $T(n)$, we must first compute $T(n - 1)$. Thus, $T(n)$ is calculated by repeated application of the formula. For example, to compute $T(6)$ we must first find $T(1), T(2), \ldots, T(6)$. Since, as we've already seen, $T(1) = 1$, it follows that

$$T(2) = 2T(1) + 1 = 3.$$

Going on like this, we have

$$T(3) = 7, \ T(4) = 15, \ T(5) = 31, \text{ and } T(6) = 63.$$

Hence to solve my puzzle, which has six discs, I have to make at least 63 moves. What about the puzzle in the temple in India? To solve it we have to compute $T(64)$, a rather dire task.

The formula $T(n) = 2T(n-1) + 1$ is an example of a *recursive formula*. In other words, to find $T(n)$ it is necessary to apply the formula several times, each time using as input the output of the previous calculation. You may be asking, How is this formula *proved*? The answer is that the dialogue above *is* a proof of the formula. Admittedly, it is couched in a form that may seem rather exotic for something that claims to be a mathematical proof. That's easy enough to cure; all you have to do is extract the main points of the proof from the dialogue and rewrite it in the usual mathematical jargon.

The fact that we obtained only a recursive formula does not stop us from dreaming of finding a closed formula for $T(n)$. That's a formula from which $T(n)$ is obtained by a simple substitution of the value of the variable n. If you look carefully at the values of $T(n)$ we computed above, you will guess what this formula ought to be. Once a closed formula has been guessed, we are faced with a new task: We must prove that it works for all values of n. Note that having guessed the formula by inspection of a table of numbers, all we can be certain of is that it holds for the numbers in the table—no more. In order to prove the formula we introduce the method of proof by *finite induction*.

You may be thinking, *Why bother to find another formula, closed or not? What's wrong with this recursive formula?* Those are reasonable questions. After all, to find $T(n)$ for a given n all we have to do is compute $T(0), \ldots, T(n)$ using the recursive formula. A computer will do that very fast. Isn't that enough? This is perhaps a point at which mathematics and computer science pull in different directions.

Exaggerating a little, we could say that computer science aims at making a brute force approach work as efficiently as possible; mathematics, on the other hand, aims at ways of reaching the goal with a minimum of calculations. Of course, these are really two faces of the same coin. In the real world, it is usually a combination of mathematics and computation that solves the problem. So the two subjects rarely compete; most often they cooperate.

2. Finite induction

The word *induction* is used in mathematics with a very special technical meaning, sometimes qualified by the adjective *finite*, as in the name of this section. But the word also has a host of other meanings, 12 of which are listed in the *Oxford English Dictionary*. The mathematical use of the word *induction* is derived from its traditional use in logic, which is also close to its everyday use. According to the *Oxford English Dictionary*, *induction* in this sense is

> the process of inferring a general law or principle from the observation of particular instances.

Therefore, when guessing the closed formula for $T(n)$ using the values of $T(1), \ldots, T(6)$, you proceeded by induction. Of course we do things like this every day, all day long. But in mathematics, induction sometimes leads to some very dreadful mistakes (the same is true in everyday life, mind you).

An often quoted example of induction gone wrong is Fermat's statement to Frenicle that all numbers of the form

$$F(n) = 2^{2^n} + 1$$

were prime. He probably checked that this was true for $n = 0, 1, 2, 3$, and 4, which is quite easy to do, and then generalized from that. The next number is

$$F(5) = 2^{2^5} + 1 = 4,294,967,297,$$

which is quite big if all you have is pen and paper. Was Fermat intimidated by the size of the number, or did he make a mistake in his calculations? We'll probably never know. But, as we have seen, $F(5)$ is really composite, and Euler was the first person to find a factor, in 1738. We shall study these numbers in detail in Chapter 9. In the meantime they stand as a warning of the dangers of generalizing from a few examples.

In the seventeenth century several mathematicians, Fermat among them, had begun to worry about the uncertainty of results obtained by induction. This led them to develop a method that is very suitable for proving results obtained by generalizing from numerical data; that is, results obtained by ordinary induction. This new method became known as *finite induction*, or *mathematical induction*, or *reasoning by recurrence*. In B. Pascal's tract *On the Arithmetical Triangle*, published in 1654, we find the method of finite induction explained in almost exactly its modern form. Of course, the "arithmetical triangle" of the title is now known as *Pascal's triangle*. Pascal was a man of many talents who did first-rate work in geometry and physics, and who invented one of the first mechanical calculating machines. His *Pensées* are a classic of French literature.

Let's go back to the tower of Hanoi for a while. Look carefully at the values of $T(1), \dots, T(6)$ computed in the previous section. You will notice that, for each of these we have $T(n) + 1 = 2^n$. Thus it is reasonable to conjecture that *the minimum number of moves required to solve a tower of Hanoi puzzle with n discs is $2^n - 1$*. Note that we have, in fact, infinitely many statements, one for each positive number. But our data tell us only that these statements are true for $n = 1, \dots, 6$—no more.

The example of the tower of Hanoi is actually fairly typical. Inspection of a (finite) table with data about a given problem often allows us to infer a statement $S(n)$ that we *expect* might be true for all positive integers n. Finite induction offers a systematic approach to proving many of these statements.

Principle of finite induction. *Suppose that, for each positive integer n, we have a statement $S(n)$ that has the following two properties:*

(1) *$S(1)$ is true.*
(2) *If $S(k)$ is true for a positive integer k, then $S(k + 1)$ is also true.*

Then $S(n)$ is true for every positive integer n.

Let's try to understand why the principle works. Suppose we have a statement for which (1) and (2) hold. Now, (2) says that if, for some integer k, we can show that $S(k)$ is true, then $S(k + 1)$ will also be true. But $S(1)$ is true by (1). So applying (2) with $k = 1$, we conclude that $S(2)$ is true. Since we now know that $S(2)$ is true, we can apply (2) again, this time with $k = 2$. This shows that $S(3)$ is true. Of course, given any positive integer n, we can carry on like this until we reach $S(n)$. Thus the statement must be true for every positive integer n.

Of course this is *not* a proof that the principle of induction works. Indeed, in a sense, this principle cannot be proved at all! Henri Poincaré, one of the greatest mathematicians of the nineteenth century, explains the point very nicely:

> The views upon which reasoning by recurrence is based may be exhibited in other forms; we may say, for instance, that in any finite collection of different integers there is always one that is smaller than any other. We may readily pass from one enunciation to another, and thus give ourselves the illusion of having proved that reasoning by recurrence is legitimate. But we shall always be brought to a full stop—we shall always come to an indemonstrable axiom, which will at bottom be but the proposition we had to prove translated in another language.

So why do we believe so strongly in the truth of the principle of induction? Let Poincaré come again to our aid:

> It is because it is only the affirmation of the power of the mind which knows it can conceive of the indefinite repetition of the same act, when the act is once possible.

It is for essentially the same reason that, being familiar only with sets with few elements, we have no difficulty in conceiving that the sequence of natural numbers is infinite. The quotations of Poincaré are from his classic book *Science and Hypothesis*; see Poincaré 1952.

Let's apply the principle to the tower of Hanoi problem. The statement we wish to prove says that the minimum number of moves $T(n)$ required to solve the tower of Hanoi puzzle with n discs is $2^n - 1$. In other words, we want to prove that $T(n) = 2^n - 1$. According to the principle of finite induction, this will be true for every integer $n \geq 1$ if we can prove two things. First, we must show that the formula holds for the puzzle with 1 disc. But we've already seen that $T(1) = 1$ and $2^1 - 1 = 1$. So there is nothing to do on this front.

Next, we must show that if the statement holds for some $k \geq 1$, then it also holds for $k + 1$. This is achieved by combining two ingredients:

- the (assumed) fact that the statement holds for some $k \geq 1$. In the example, this means we are supposing that $T(k) = 2^k - 1$ for some $k \geq 1$. This is called the *induction hypothesis*.
- some relation between $T(k)$ and $T(k+1)$. In the example this is provided by the recursive relation $T(k + 1) = 2T(k) + 1$.

Thus, suppose that $T(k) = 2^k - 1$ for some $k \geq 1$. It follows from the recursive relation that

$$T(k + 1) = 2T(k) + 1 = 2(2^k - 1) + 1 = 2^{k+1} - 2 + 1 = 2^{k+1} - 1.$$

Thus condition (2) of the principle of finite induction also holds in this case. Now the principle of induction takes over: Since (1) and (2) hold, then $T(n) = 2^n - 1$ for every $n \geq 1$.

Having proved the formula for the minimum number of moves to solve the tower of Hanoi puzzle, we can now find out how much time we have left before the world ends. Recall that the primordial tower had 64 discs. Thus the total number of moves the priests have to make from the day of creation to the end of time is $T(64) = 2^{64} - 1$. But what we want to know is how long it will take to move the discs this many times. Let's assume that the priests need, on average, 30 minutes to move a disc. The discs are of different sizes, but we haven't been told how big they are. Quite big, probably, since they're the work of God. If they're made of gold, they'll be very heavy. So 30 minutes is really a conservative estimate. Since 2^{64} is of the order of magnitude of 10^{19}, a simple calculation shows that the priests will take approximately 10^{14} years to move all the discs. The latest calculations show that 10^{11} years have passed since the Big Bang, which leaves us with plenty of time.

This legend was first published in Paris in 1883, at the same time as the puzzle, by a certain N. Claus de Siam of the College of Li-Sou-Stian. The names of the man and the college are really anagrams of Lucas d'Amiens, who taught at the Lycée Saint-Louis. This was the mathematician F. E. A. Lucas, who invented both the puzzle and the tale about its origin. Lucas's book *Récréations mathématiques*, of 1894, became a classic on the subject. Lucas also worked in number theory, and in Chapters 9 and 10 we will study two primality tests he discovered. Using one of his tests, Lucas showed, without the aid of a computer, that the Mersenne number

$$M(127) = 170{,}141{,}183{,}460{,}469{,}231{,}731{,}687{,}303{,}715{,}884{,}105{,}727$$

is prime.

3. Fermat's theorem

The theorem we want to prove is sometimes affectionately called "Fermat's Little Theorem". It says that if p is a prime number and a is any integer, then p divides $a^p - a$. Some special cases of this result had been known for hundreds of years, but Fermat seems to have been the first to state the result in full generality. Let's begin by translating the theorem in terms of congruences.

Fermat's theorem. *Let $p > 0$ be a prime number and let a be an integer; then*

$$a^p \equiv a \pmod{p}.$$

To prove this theorem using finite induction we must find a proposition $P(n)$ to which the method can be applied. The proposition is

$$n^p \equiv n \pmod{p} \quad \text{for an integer} \quad n.$$

Note that the proposition says only that the congruence of the theorem holds for positive integers. Thus, apparently, we are not proving the theorem in full generality. However, every integer is congruent, modulo p, to a non-negative integer smaller than p. Hence, it is enough to prove the theorem for $0 \leq a \leq$

$p - 1$. In particular, the theorem follows if we prove that the proposition $P(n)$ holds for every $n \geq 1$.

Of course $P(1)$ holds, since $1^p = 1$. To go from $P(n)$ to $P(n + 1)$ we must find a way to relate these two statements. This is provided by the *binomial theorem*. The proof becomes considerably simpler if we isolate as an auxiliary result what is really a version of the binomial theorem for integers modulo p.

Lemma. *Let $p > 0$ be a prime number and let a and b be integers; then*

$$(a + b)^p \equiv a^p + b^p \pmod{p}.$$

Proof of the lemma. It follows from the binomial formula that

$$(a + b)^p = a^p + b^p + \sum_{i=1}^{p-1} \binom{p}{i} a^{p-i} b^i.$$

Thus, to prove the lemma, it is enough to show that

$$\sum_{i=1}^{p-1} \binom{p}{i} a^{p-i} b^i \equiv 0 \pmod{p}.$$

But this follows immediately if we prove that the binomial numbers $\binom{p}{i}$ are divisible by p for $1 \leq i \leq p - 1$. Now, by definition

$$\binom{p}{i} = \frac{p(p-1)\ldots(p-i+1)}{i!}.$$

Since binomial numbers are integers, the denominator of this fraction must divide the numerator. But if $1 \leq i \leq p - 1$, then p is *not* a factor of $i!$ Thus the factor p that appears in the numerator of the fraction will *not* be canceled by the denominator. Therefore $i!$ must divide $(p-1)\ldots(p-i+1)$ and we have that $\binom{p}{i}$ is a multiple of p, as we wanted to prove.

We may now return to the proof of Fermat's theorem. The induction hypothesis is

$$n^p \equiv n \pmod{p} \quad \text{for some integer } n,$$

and we must show that $(n + 1)^p \equiv n + 1 \pmod{p}$. By the lemma

$$(n + 1)^p \equiv n^p + 1^p \equiv n^p + 1 \pmod{p}.$$

By the induction hypothesis we can replace n^p by n in this formula. Doing this, we conclude that

$$(n + 1)^p \equiv n^p + 1 \equiv n + 1 \pmod{p},$$

as we wanted to prove.

The more interesting applications of Fermat's theorem will have to wait until the next chapter. For now, we will be content with using the theorem to simplify the computation of powers modulo p, a problem that we have already worked on. First, we must state the theorem in a more convenient form.

According to the theorem, if p is a prime and a is any integer, then $a^p \equiv a$ (mod p). Suppose now that the integer is *not* divisible by p. Since p is a prime number, a and p are then co-prime. Thus it follows from the invertibility theorem that a is invertible modulo p; let a' be its inverse. Multiplying $a^p \equiv a$ (mod p) by a', we have

$$a'a \cdot a^{p-1} \equiv a'a \quad (\text{mod } p).$$

But $a'a \equiv 1$ (mod p), so we end up with $a^{p-1} \equiv 1$ (mod p). This is the version of the equation of Fermat's theorem we will use most often. Let's state it here for future reference.

Fermat's theorem. *Let p be a prime number and let a be an integer that is not divisible by p. Then $a^{p-1} \equiv 1$ (mod p).*

The problem to which we want to apply Fermat's theorem is the following. Given three positive integers a, k, and p, such that $k > p - 1$, find the residue of a^k modulo p.

If p is a factor of a, the residue is 0. So we may assume that p is not a factor of a. Now divide k by $p - 1$ to get $k = (p-1)q + r$ where q and r are non-negative integers and $0 \le r < p - 1$. Therefore

$$a^k \equiv a^{(p-1)q+r} \equiv (a^{p-1})^q a^r \quad (\text{mod } p).$$

But $a^{p-1} \equiv 1$ (mod p) by Fermat's theorem. Hence $a^k \equiv a^r$ (mod p). Thus the calculation can proceed with an exponent that is smaller than $p - 1$.

Here is a dramatic example of the power of this simple reduction. Suppose we want to find the residue of $2^{5,432,675}$ modulo 13. The procedure of the previous chapter would force us to calculate several powers of 2 modulo 13 before we got anywhere. Let's see what happens if we use Fermat's theorem. First find the remainder of the division of 5,432,675 by $13 - 1 = 12$, which is 11. Thus, arguing as above, we get

$$2^{5,432,675} \equiv 2^{11} \quad (\text{mod } 13).$$

A straightforward calculation shows that $2^{11} \equiv 7$ (mod 13).

4. Counting roots

There is still one question we must consider. Is there a positive integer k, *smaller* than $p - 1$, for which $a^k \equiv 1$ (mod p) for *every* integer a that is not divisible by p?

One might try to answer this question as follows. There is a well-known theorem that says a polynomial equation cannot have more roots than its degree. But since $a^k \equiv 1$ (mod p) holds for every integer a co-prime to p, it follows that the roots of the polynomial equation $x^k = \bar{1}$ in $\mathbb{Z}_p$ are $\bar{1}, \ldots, \overline{p-1}$. Therefore the equation has $p - 1$ distinct roots. By the theorem just quoted this means that $k \ge p - 1$. So the answer to the question posed above must be no.

Although the argument of the previous paragraph is correct, it hides the true nature of the problem, which lies with the theorem that has been used to settle the question. When we deal with polynomial equations, it usually means an equation with real or complex coefficients. And when we say "root", we mean real or complex root. But the coefficients of the polynomial equation above, and its roots, are elements of $\mathbb{Z}_p$. Therein lies the problem. We must show that the theorem that relates the number of roots to the degree remains valid in this case. Is this just another case of unjustified pedantry? Not at all. Although the theorem is indeed true when the modulus is prime, it is *false* for a composite modulus.

Theorem. *Let $f(x)$ be a polynomial of degree k with integer coefficients and leading coefficient 1. If p is a prime number, then $f(x)$ cannot have more than k distinct roots in $\mathbb{Z}_p$.*

Before we plunge into the proof, there are two points we must deal with. First, the modulus must be prime because we need to know that if $ab \equiv 0$ (mod p), then $a \equiv 0$ (mod p) or $b \equiv 0$ (mod p). Note that this is actually the fundamental property of prime numbers rephrased in the language of congruences. Second, we will isolate one of the basic ingredients of the proof in a lemma that will be proved at the end of the section.

Lemma. *Let $h(x)$ be a polynomial with integer coefficients and degree m. Given an integer α, there exists a polynomial $q(x)$ of degree $m - 1$ such that*

$$h(x) = (x - \alpha)q(x) + h(\alpha).$$

The proof of the theorem is by induction on the degree n of the polynomial f—with a little help from the lemma. If $n = 1$, then $f(x) = x + b$. So the polynomial has only the solution $\overline{-b}$ in $\mathbb{Z}_p$. Thus, a polynomial of degree 1 has only one solution in $\mathbb{Z}_p$, and the theorem is proved in this case.

Suppose now that *every* polynomial of degree $k - 1$, whose leading coefficient is 1, has at most $k-1$ distinct roots in $\mathbb{Z}_p$. This is the induction hypothesis. We wish to show that this implies the same holds for a polynomial of degree k and leading coefficient 1.

Thus let $f(x)$ be a polynomial of degree k, with integer coefficients, and leading coefficient 1. If $f(x)$ has no roots in $\mathbb{Z}_p$, we are done, for $0 \le k$. Such polynomials exist; an example is given at the end of the proof. Therefore we may assume that $f(x)$ has a root $\overline{\alpha} \in \mathbb{Z}_p$; in other words, $f(\alpha) \equiv 0$ (mod p). By the lemma

(4.1) $f(x) = (x - \alpha)q(x) + f(\alpha),$

where $q(x)$ has degree $k - 1$. Note that since both $f(x)$ and $x - \alpha$ have leading coefficient 1, this is also true of $q(x)$. Hence we can apply the induction hypothesis to $q(x)$.

Now, reducing equation (4.1) modulo p, we get

(4.2) $f(x) \equiv (x - \alpha)q(x) \pmod{p}.$

Let $\overline{\beta} \neq \overline{\alpha}$ be another root of $f(x)$ in $\mathbb{Z}_p$. This means that

$$f(\beta) \equiv 0 \pmod{p} \quad \text{but} \quad \alpha - \beta \not\equiv 0 \pmod{p}.$$

Replacing x by α in (4.2) and using these congruences, we have

$$0 \equiv f(\beta) \equiv (\beta - \alpha)q(\beta) \pmod{p}.$$

Since p is prime, this implies that $q(\beta) \equiv 0 \pmod{p}$. We conclude that if $\overline{\beta}$ is a root of $f(x)$ in $\mathbb{Z}_p$, distinct from $\overline{\alpha}$, then $\overline{\beta}$ is a root of $q(x)$ in $\mathbb{Z}_p$. In other words, $f(x)$ can have only one more root (in $\mathbb{Z}_p$) than $q(x)$. But by the induction hypothesis, $q(x)$ has at most $k - 1$ distinct roots in $\mathbb{Z}_p$. Hence $f(x)$ cannot have more than k distinct roots, and the proof by induction is complete.

Let's consider some examples. The first is the polynomial $f(x) = x^2 + 3$. It satisfies all the conditions of the theorem, but has no roots modulo 5. Indeed, the only possible residues for the square of any integer modulo 5 are 1 and 4. This is the example we mentioned in the proof of the theorem.

The second example illustrates what happens when we try to find the roots of a polynomial under a composite modulus. For example, the roots of the polynomial $x^2 - 170$ in $\mathbb{Z}_{385}$ are $\overline{95}$, $\overline{150}$, $\overline{235}$, and $\overline{290}$, as one easily checks. Thus we have a polynomial equation of degree 2 with 4 roots. None of this contradicts the theorem, of course, because 385 is a composite number.

In order to complete the proof of the theorem we must prove the lemma. This we will do, once again, by induction. However, if you try to apply the principle of induction in the form of section 2, you'll realize that a gap opens in the proof. We'll see what the problem is as we go through the proof.

We get over this problem by stating the principle of induction in a slightly different form. The induction hypothesis of the principle in the form of section 2 assumes that $S(k)$ is true for some integer $k \geq 1$. In practice, if we are trying to prove $S(k+1)$ from $S(k)$, then we already know that $S(1), S(2), \ldots, S(k)$ are true. Thus little will change if we assume in the induction hypothesis that not only $S(k)$, but also $S(1), S(2), \ldots, S(k-1)$ are true, and use all this information to prove $S(k+1)$. Incorporating this change, the statement of the principle is the following.

Principle of finite induction. *Suppose that, for each positive integer n, we have a statement $S(n)$ that has the following two properties:*

(1) *$S(1)$ is true; and*
(2) *If $S(1), \ldots, S(k)$ are true for a positive integer k, then $S(k+1)$ is also true.*

Then $S(n)$ is true for every positive integer n.

Using this form of the principle, it is now easy to prove the lemma by induction on the degree m of $h(x)$. If $m = 1$, there exist integers a and b such that $h(x) = ax + b$. Hence

$$h(x) = ax + b = a(x - \alpha) + a\alpha + b = a(x - \alpha) + h(\alpha).$$

Suppose now that the lemma holds for every polynomial with integer coefficients and degree less than or equal to $m - 1$. From this fact we wish to conclude that the lemma holds for a polynomial $h(x)$ with integer coefficients and degree m. Suppose that

$$h(x) = a_m x^m + a_{m-1} x^{m-1} + \cdots + a_1 x + a_0,$$

where $a_m \neq 0$. Let $g(x) = h(x) - a_m x^{m-1}(x - \alpha)$; that is,

$$g(x) = (a_{m-1} - a_m \alpha) x^{m-1} + a_{m-2} x^{m-2} + \cdots + a_1 x + a_0.$$

The degree of $g(x)$ is clearly less than or equal to $m - 1$. Note, however, that the degree will be exactly $m - 1$ only if $a_{m-1} - a_m \alpha \neq 0$, but we have no way of knowing whether or not this is the case. Thus it is possible for $g(x)$ to have degree *less* than $m - 1$. This is where we would run into trouble if we were using the principle of induction in the form of section 2. Note also, for future reference, that $g(\alpha) = h(\alpha)$.

Since $g(x)$ has degree less than or equal to $m-1$, it follows by the induction hypothesis that

$$g(x) = j(x)(x - \alpha) + g(\alpha),$$

where $j(x)$ is a polynomial with integer coefficients and degree one less than the degree of $g(x)$. Since $g(\alpha) = h(\alpha)$, we have

$$g(x) = j(x)(x - \alpha) + h(\alpha).$$

But $h(x) = g(x) + a_m x^{m-1}(x - \alpha)$, so

$$h(x) = (j(x) + a_m x^{m-1})(x - \alpha) + h(\alpha).$$

Finally, $j(x) + a_m x^{m-1}$ has degree exactly $m - 1$ because the degree of $j(x)$ is less than $m - 1$. This completes the proof of the lemma.

There is a more direct proof of the lemma, but it assumes prior knowledge of the division of polynomials. As before, let $h(x)$ be a polynomial of degree m and integer coefficients. Dividing $h(x)$ by $x - \alpha$, we find two polynomials $q(x)$ and $r(x)$ such that

(4.3) $$h(x) = q(x)(x - \alpha) + r(x),$$

and $r(x) = 0$ or $r(x)$ has degree less than $x - \alpha$. Thus $r(x)$ must have degree 0, which means that $r(x) = c$ is an integer. Replacing x by α in equation (4.3),

$$h(\alpha) = q(\alpha)(\alpha - \alpha) + c = c.$$

Hence we can rewrite (4.3) in the form

$$h(x) = q(x)(x - \alpha) + h(\alpha),$$

and the proof is complete.

5. Exercises

1. Prove by induction that:

(1) $n^3 + 2n$ is divisible by 3 for all integers $n \geq 1$.

(2) If $n > 0$ is an odd integer, then $n^3 - n$ is always divisible by 24.

(3) A convex polygon of n sides has exactly $n(n - 3)/2$ diagonals.

(4) For each integer $n \geq 1$, $\sum_{k=1}^{n} k(k + 1) = n(n + 1)(n + 2)/3$.

2. The numbers defined by the formula $h_n = 1 + 3n(n - 1)$, for $n = 1, 2, \ldots$, are called *hexagonal*. The name comes from the fact that these numbers can be arranged around the sides of concentric regular hexagons.

(1) Compute the sum of the first n hexagonal numbers when $n = 1, 2, 3, 4$, and 5. Use these data to guess a formula for the sum of the first n hexagonal numbers.

(2) Prove the formula obtained in (1) by finite induction.

3. Recall that if f_n is the nth number in the Fibonacci sequence, then $f_0 = f_1 = 1$ and $f_n = f_{n-1} + f_{n-2}$. Show by induction on n that

$$f_n = \frac{\alpha^n - \beta^n}{\sqrt{5}},$$

where α and β are the roots of the quadratic equation $x^2 - x - 1 = 0$.

4. Consider the sequence of positive integers $S_0, S_1, S_2 \ldots$, defined recursively by

$$S_0 = 4 \quad \text{and} \quad S_{k+1} = S_k^2 - 2.$$

Let $\omega = 2 + \sqrt{3}$ and $\varpi = 2 - \sqrt{3}$. Show by induction on n that $S_n = \omega^{2^n} + \varpi^{2^n}$.

5. The principle of finite induction is very useful, but it has to be handled with care. For example, what is wrong with the following proof that, in a finite set of colored balls, all the balls must have the same color? The proof is by induction on the number of elements of the set.

> If the set has only one ball, then the statement is clearly true. Now assume that in every set with k colored balls, all the balls have the same color. We want to use this to show that all the balls in a set with $k + 1$ balls must have the same color. Denote the balls in this latter set by $b_1, \ldots, b_{k+1}$. Removing b_{k+1}, we have a set with k balls. Thus, by the induction hypothesis, $b_1, \ldots, b_k$ have the same color. So we need only show that b_{k+1} has the same color as one of the balls in the set $\{b_1, \ldots, b_k\}$, and the proof will be complete. But the set $b_2, \ldots, b_{k+1}$ also has k elements, so all its balls have the same color. Thus b_{k+1} has the same color as, say, b_2. Hence the balls $b_1, \ldots, b_{k+1}$ have the same color.

6. Suppose we have 3^n coins, one of which is false and weighs less than the other ones. You are given a pair of scales, but no weights. The only way to weigh the coins is to put some on one scale, some on the other, and check whether they are balanced. Show, by finite induction, that n trials are enough to find the false coin.

7. Let p_n be the nth prime number. Thus, for example, $p_1 = 2$, $p_2 = 3$, and $p_3 = 5$. We want to find an upper bound for the nth prime in terms of n.

(1) Show that $p_{n+1} \leq p_1 \ldots p_n + 1$.
(2) Using finite induction and (1), show that the nth prime satisfies the inequality $p_n \leq 2^{2^n}$.

8. Show, using Fermat's theorem, that $2^{70} + 3^{70}$ is divisible by 13.

9. Let a be a positive integer written in basis 10. Show that the unit digits of a^5 and a are the same.

10. Use Fermat's theorem to show that for all integers n, the number $n^3 + (n+1)^3 + (n+2)^3$ is divisible by 9.

11. Let $p \neq 2, 5$ be a prime number. Show that p divides a member of the set
$$\{1; 11; 111; 1111; 11,111; \ldots\}.$$
Hint: By Fermat's theorem, we have that if $p > 5$, then $10^{p-1} - 1$ is divisible by p. The case $p = 3$ must be dealt with separately.

12. Show that the equation $x^{13} + 12x + 13y^6 = 1$ has no integer solutions.
Hint: Reduce modulo 13 and use Fermat's theorem.

13. Find the remainder of the division of
(1) $39^{50!}$ by 2251.
(2) 19^{39^4} by 191.

14. The purpose of this problem is to show that if $p = 4n + 1$ is prime, then there are integers a and b such that p divides $a^2 + b^2$. Let x and y be two integers prime with p. Put $a = x^n$ and $b = y^n$. Then
$$(a^2 - b^2)(a^2 + b^2) = x^{4n} - y^{4n}.$$
(1) Use Fermat's theorem to show that $x^{4n} - y^{4n}$ is divisible by p.
(2) Use (1) to show that p divides either $a^2 + b^2$ or $a^2 - b^2$.
If p divides $a^2 + b^2$, the result is proved. Thus we can assume, by contradiction, that for any integers x and y, we have that p divides $x^{2n} - y^{2n} = a^2 - b^2$. In particular, this must hold when x is any integer and $y = 1$. In this case, $x^{2n} - 1$ is divisible by p for any integer x.
(3) Show that if $x^{2n} \equiv 1 \pmod{p}$ for every integer x, then we have a contradiction with the theorem of section 4.
(4) Use these results to show that there must exist integers a and b such that $a^2 + b^2$ is divisible by p.

Fermat knew the stronger result that every prime of the form $4n + 1$ can be written as a sum of two squares. See Weil 1987, Chapter II, sections VII and VIII. Compare the above result with exercise 13 of Chapter 4.

15. In this exercise we describe Euler's proof of Fermat's theorem. Unlike Fermat's proof, this one does not use finite induction. Let p be a prime and $\overline{a} \neq \overline{0}$ be an element of $U(p) = \mathbb{Z}_p \setminus \{\overline{0}\}$. Consider the subset
$$S = \{\overline{a}, \overline{2a}, \ldots, \overline{(p-1)a}\}.$$
(1) Show that the elements of S are all distinct.
(2) Show that S has $p - 1$ elements, and conclude that $S = U(p)$.

(3) Show that (2) implies that the product of all the elements of S is equal to $\overline{(p-1)!} = \overline{1} \cdot \overline{2} \cdots \overline{(p-1)}$.

(4) Show that the product of the elements of S is also equal to $\overline{a}^{p-1}\overline{(p-1)!}$.

(5) Show that (3) and (4) imply Fermat's theorem.

16. Let p be a prime number and let a be an integer that is not divisible by p. Show that the inverse of $\overline{a}$ in $\mathbb{Z}_p$ is $\overline{a}^{p-2}$.

17. Let $p = 4k+3$ be a positive prime. Given an integer a, consider the equation $x^2 \equiv a$ (mod p).

(1) Give examples of a and p for which the equation does not have a solution.

(2) Show that if the equation has a solution, it is congruent modulo p to $\pm a^{k+1}$.

Hint: If the equation has a solution, then there is an integer b such that $b^2 \equiv a$ (mod p). Thus

$$(a^{k+1})^2 \equiv b^{4(k+1)} \equiv b^{4k+2} \cdot b^2 \pmod{p}.$$

Now (2) follows immediately by Fermat's theorem.

18. Use your work from exercise 16 to write a program that, given an integer a and a prime p as inputs, finds the inverse of a modulo p. The program should check, first of all, whether p divides a.

19. Let $p = 4k + 3$ be a positive prime. Write a program that, given p and a positive integer a, computes the solutions of $x^2 \equiv a$ (mod p). Note that we know from exercise 17 that if this equation has a solution b, then $b \equiv \pm a^{k+1}$ (mod p). Thus the program will compute the residue of a^{k+1} modulo p and then check whether this number is really a solution of the given equation. The output will be either the solution of the equation or a message that it does not have a solution. This is the second exercise of a sequence that ends with exercise 8 of Chapter 11.

20. Some problems in the theory of numbers lead to primes that satisfy the equation

$$a^{p-1} \equiv 1 \pmod{p^2}$$

for some integer a. Write a program that, having positive integers $a > 1$ and r as input, finds all the primes p in the interval $a + 1$ to r that satisfy the above equation. The program must first use the sieve of Erathostenes to find the primes smaller than or equal to r. These primes are then subject to a search for those that also satisfy the congruence above. If $r = 10^5$, then the number of primes satisfying the congruence is two when $a = 2, 5, 10$, and 14; and five when $a = 19$.

6

Pseudoprimes

In this chapter we will see how Fermat's theorem can be used to show that a number is composite in a way that avoids all searches for factors. The chapter closes with a discussion of the strategies that various computer algebra systems use to check whether a number is prime or composite.

1. Pseudoprimes

It follows from Fermat's theorem that if p is a prime number, and a is an integer not divisible by p, then $a^{p-1} \equiv 1 \pmod{p}$. Now suppose that we are asked whether a certain positive *odd* integer n is prime. Suppose also that somehow we manage to find an integer b, which is not divisible by p, and for which $b^{n-1} \not\equiv 1 \pmod{n}$. Then it follows from Fermat's theorem that n cannot be prime. The number b is called a *witness* to the fact that n is composite. Thus we have a method for testing compositeness that does not depend on factoring the number. The problem with this test is that it will not work unless we are capable of finding a witness; and for that one needs luck. But, as we shall see, it is more likely that such a witness will be found than not.

Note that we need not look for a witness b among all the integers. Indeed, since we are working with congruences modulo n, it is possible to restrict the search to the interval $0 \leq b \leq n - 1$. We can also exclude 0, because b must not be a multiple of n, and 1, because $1^{n-1} \equiv 1 \pmod{n}$ holds for every n. Moreover, since we are assuming that n is odd, it follows that $(n-1)^{n-1} \equiv 1 \pmod{n}$. Thus the congruence is also satisfied by $n - 1$. Therefore we can assume that $1 < b < n - 1$. Before we put the test to use, let's state it in the form of a theorem.

Compositeness test. *Let $n > 0$ be an odd integer. If there exists an integer b such that*

(1) $1 < b < n - 1$, *and*
(2) $b^{n-1} \not\equiv 1 \pmod{n}$,

then n is composite.

Recall that the rep-unit $R(n)$ is defined by the formula

$$R(n) = \frac{10^n - 1}{9}.$$

In other words, it is the integer obtained by repeating n times the digit 1; see exercise 5 of Chapter 2. We have seen that $R(n)$ is composite, if n is composite.

However, since 229 is prime, we have no way to tell in advance whether $R(229)$ is prime or composite. Moreover, the number has more than 200 digits, so it is not a good idea to use a factorization algorithm. Instead, let's apply the compositeness test with $b = 2$. With the help of a computer algebra system it is easy to compute the residue of $2^{R(229)-1}$ modulo $R(229)$; it is

10451650058433339778175376888598283548861273723388489857084
82884056668984062908255365523134523742682565391455276061215
67512885287283062854774198632697829520351103663852079821692
41234610147904074388417006924857636593110454503 29217,

which is not congruent to 1 modulo $R(229)$. Therefore $R(229)$ is composite.

Since we are more interested in prime numbers than in composite numbers, it is reasonable to ask whether Fermat's theorem can be used to prove that a number is prime. More precisely, suppose that $n > 0$ is an odd integer that satisfies $b^{n-1} \equiv 1 \pmod{n}$ for *some* integer $1 < b < n - 1$; is n necessarily prime? Leibniz, the famous philosopher and mathematician, believed that the answer was affirmative. He actually used this as a primality test, always choosing $b = 2$, to simplify the calculations.

Unfortunately, Leibniz was quite wrong in this respect. For example, $2^{340} \equiv 1 \pmod{341}$. Thus, according to Leibniz, 341 ought to be prime. But $341 = 11 \cdot 31$ is composite. The numbers that give a "false positive" result in this test are known as pseudoprimes. In other words, if a positive integer n is odd and composite and satisfies $b^{n-1} \equiv 1 \pmod{n}$, for some integer $1 < b < n - 1$, then it is called a *pseudoprime* to the *base* b. Therefore, 341 is a pseudoprime to base 2.

While certainly not 100 percent accurate, Leibniz's test is still quite useful. For small numbers chosen randomly, the test wins far more times than it loses. To see why this is so, let's count the primes and the pseudoprimes to base 2, up to a certain bound. For example, between 1 and 10^9 there are 50,847,534 primes, but only 5597 pseudoprimes to base 2. Thus a number in this range that passes Leibniz's test is far more likely to be prime than to be a pseudoprime to base 2.

Moreover, we have been applying the test to only one base, but if several different bases are used, then the number of composites that go undetected is considerably reduced. For example, $3^{340} \equiv 56 \pmod{341}$, so that 3 is a witness to the fact that 341 is composite. Indeed, between 1 and 10^9 there are 1272 pseudoprimes to bases 2 and 3, and only 685 pseudoprimes to bases 2, 3, and 5.

Since we need to test only for finitely many bases, it is reasonable to ask whether n can be composite and yet a pseudoprime to all these bases. Let $n > 2$ and suppose that $b^{n-1} \equiv 1 \pmod{n}$ for some $1 < b < n - 1$. Since $b^{n-1} = b \cdot b^{n-2}$ it follows from the congruence that b is invertible modulo n. By the invertibility theorem this is equivalent to saying that $\gcd(b, n) = 1$. Thus, if n is composite, and b is not prime to n, then $b^{n-1} \not\equiv 1 \pmod{n}$. In particular, any factor of n will be a witness that n is composite. Thus the answer to the question above must be no.

What can we conclude, in light of the result of the previous paragraph? Recall that our aim is to find an efficient way to determine whether a given number is prime. Now, if the number has a small factor, it is easily found with the trial division algorithm of Chapter 2. So, in practice, the compositeness test will be applied only if the factors of the number under consideration are all fairly large. Thus the question asked above is of little practical interest. It would be faster to search for a factor than to try to check whether a big number n is a pseudoprime to all bases between 2 and $n - 2$. That's not the end of the story, though, as we will see in the next section.

Before we move on, it may be advisable to note that several books have a value for $\pi(10^9)$, the number of positive primes smaller than 10^9, that is smaller than the one given above. These are not random typographical errors, because all the books give the same number. What happened was that in 1893 the Danish mathematician N. P. Bertelsen made a mistake in his calculation of $\pi(10^9)$, as a consequence of which his count was 56 primes short of the correct value. Ironically, his calculation was meant to correct errors in tables. Instead, he introduced an error that can be found in books published as recently as 1993.

2. Carmichael numbers

As we saw at the end of the last section, if a positive odd integer n is composite and a base b is chosen that is not prime to n, then n will not be a pseudoprime to base b. Unfortunately, this is not very helpful. In practice, to keep the calculations within bounds, we just select a few bases amongst the smaller primes. If the smallest factor of n is very large, all these bases will be prime to n. So the question one really ought to ask is an improved form of the one dealt with at the end of section 1. The new question is, Can an odd positive integer be a pseudoprime to all bases b *that are prime to* n? Predictably, the answer to this question is yes.

Note that if b is prime to n, then $b^n \equiv b \pmod{n}$ implies that $b^{n-1} \equiv 1 \pmod{n}$. Thus we are led to ask the (nominally stronger) question: Are there odd positive integers n that are *composite*, and yet satisfy $b^n \equiv b \pmod{n}$ for all integers b? One of the advantages of rewriting the question in this form is that it does not require any extra hypothesis on b. The first mathematician to give examples of such numbers was R. D. Carmichael in a paper published in 1912; see Carmichael 1912. That's why they are called Carmichael numbers.

Since these numbers play a very important role in much of what we have to say, it is a good idea to give a formal definition. An odd integer $n > 0$ is a *Carmichael number* if n is composite and $b^n \equiv b \pmod{n}$ for all integers b. Of course, we need only check the congruence for $1 < b < n - 1$, since we are working modulo n.

As Carmichael himself showed, the smallest Carmichael number is 561. In principle we can check this from the definition. However, even for a relatively small number, this is very long and tedious. Note that to prove that 561 is a Carmichael number *directly from the definition*, we have to check that the congruence $b^{561} \equiv b$ (mod 561) holds for $b = 2, 3, 4, \ldots, 559$, which gives a total of 557 bases. This may not seem like much work if you have a computer, but how about showing that

$$349{,}407{,}515{,}342{,}287{,}435{,}050{,}603{,}204{,}719{,}587{,}201$$

is a Carmichael number? It is surely time to return to the theoretical drawing board.

Let's try an indirect approach to proving that 561 is a Carmichael number. First, note that this number is easily factored:

$$561 = 3 \cdot 11 \cdot 17.$$

Now let b be an integer; we wish to show that

(2.1) $$b^{561} \equiv b \quad (\text{mod } 561).$$

Our strategy consists of showing that $b^{561} - b$ is divisible by 3, by 11, and by 17. Since these are *distinct* primes, we can use the lemma of Chapter 2, section 6, to conclude that the product of these primes divides $b^{561} - b$. But the product is 561; so we have proved that (2.1) holds.

To make this strategy work we must be able to prove that $b^{561} - b$ is divisible by each of the prime factors of 561. That's where Fermat's theorem comes to our aid. We will do the calculations in detail for 17, and leave 3 and 11 as exercises. So we want to show that $b^{561} - b$ is divisible by 17, or, using congruences, that

(2.2) $$b^{561} \equiv b \quad (\text{mod } 17).$$

There are two cases to be considered. If 17 divides b, then both sides of (2.2) are congruent to zero modulo 17. So the congruence is verified in this case. Assume now that 17 does not divide b; then it follows from Fermat's theorem that $b^{16} \equiv 1$ (mod 17). Before we apply this to (2.2) we must determine the remainder of the division of 561 by 16. But $561 = 35 \cdot 16 + 1$. Therefore

$$b^{561} \equiv (b^{16})^{35} \cdot b \equiv b \quad (\text{mod } 17).$$

Note that the calculation followed so easily from Fermat's theorem because the remainder of the division of 561 by 16 was 1. Luckily, the remainders of the divisions of 561 by $2(= 3 - 1)$, and by $10(= 11 - 1)$ are also 1. So the calculations for 3 and 11 are as straightforward as the one above.

Thus the success of this strategy depended on two properties of 561. First, dividing 561 by each of its prime factors minus 1 left remainder 1. Second, each prime factor appears with multiplicity one in the factorization of 561. The evidence seems to point to the fact that we were very lucky in our choice of example. Either that, or Carmichael numbers must be very rare. The truth turns out to be very surprising. There are infinitely many Carmichael numbers, and they all share the properties that made the calculations with 561 so easy to perform. The characterization of Carmichael numbers that follows from these observations was first given by A. Korselt thirteen years before Carmichael's paper on the subject. However, Korselt never gave any examples of numbers that satisfied the properties he enunciated.

Korselt's theorem. *An odd integer $n > 0$ is a Carmichael number if and only if the following two conditions hold for each prime factor p of n:*

(1) p^2 *does not divide n; and*

(2) $p - 1$ *divides $n - 1$.*

Let's show first that if a number n satisfies (1) and (2) above, then it is a Carmichael number. To do this we repeat the strategy used to show that 561 is a Carmichael number. Suppose that p is a prime factor of n. We show first that

(2.3) $b^n \equiv b \pmod{p}$.

If b is divisible by p, then both sides of (2.3) are congruent to zero. Hence the congruence holds in this case. Assume now that p does not divide b. It follows from Fermat's theorem that $b^{p-1} \equiv 1 \pmod{p}$. Before we apply this to (2.3), we must determine the remainder of the division of n by $p-1$. But $p-1$ divides $n-1$ by (2). Thus $n - 1 = (p - 1)q$, for some integer q, and

$$n = (n - 1) + 1 = (p - 1)q + 1.$$

Therefore

$$b^n \equiv (b^{p-1})^q \cdot b \equiv b \pmod{p},$$

where the second congruence follows from Fermat's theorem. Summing up, if p is a prime factor of n, then $b^n \equiv b \pmod{p}$ for every integer b.

Now it follows from (1) that $n = p_1 \ldots p_k$, where $p_1 < \cdots < p_k$ are prime numbers. But we've already seen that $b^n - b$ is divisible by each one of these primes. Since all the primes are distinct, it follows from the lemma of Chapter 2, section 6, that $b^n - b$ is divisible by $p_1 \ldots p_k = n$. In other words, $b^n \equiv b$ (mod n). Since these calculations hold for any integer b, we conclude that n is a Carmichael number.

We have shown that if n satisfies (1) and (2), then it is a Carmichael number. Let's now turn to the converse. Suppose that n is a Carmichael number. Let's prove first that if p^2 divides n, then we obtain a contradiction. This will show that (1) holds if n is a Carmichael number.

Note that, since n is a Carmichael number, the contradiction will arise if we find an integer b such that $b^n \not\equiv b \pmod{n}$. Choose $b = p$. Then

$$p^n - p = p(p^{n-1} - 1).$$

But p does not divide $p^{n-1} - 1$, so p^2 cannot divide $p^n - p$. Since we are assuming that p^2 divides n, it follows that n cannot divide $p^n - p$. In other words, $p^n \not\equiv p \pmod{n}$, and n is not a Carmichael number.

To complete the proof we have only to show that if n is a Carmichael number, then (2) holds. However, this requires the *primitive root theorem*, which will be proved only in Chapter 10. The end of the proof of Korselt's theorem is therefore postponed to Chapter 10, section 3.

Unfortunately, to check that a given integer is a Carmichael number using Korselt's theorem, we must first find its complete factorization into primes. If the number is large, this can be a very daunting task. Luckily, it is often the case that very large Carmichael numbers have many small factors. For example, the number with 36 digits given at the beginning of the section is the smallest Carmichael number with 20 prime factors. Its complete factorization is

$$11 \cdot 13 \cdot 17 \cdot 19 \cdot 29 \cdot 31 \cdot 37 \cdot 41 \cdot 43 \cdot 61 \cdot 71 \cdot 73 \cdot 97 \cdot 101 \cdot 109 \cdot 113 \cdot 151 \cdot 181 \cdot 193 \cdot 641.$$

Using a computer algebra system, one now quickly checks that this is a Carmichael number. See exercise 3 for a family of integers that contains several Carmichael numbers.

In his paper of 1912, Carmichael gave fifteen examples of the numbers that now take his name, and then added, *"this list can be continued indefinitely"*. Thus he seemed to be implicitly saying that there were infinitely many Carmichael numbers. However, it soon became clear that this would be very difficult to prove. The reason why the problem is difficult is that Carmichael numbers are so very rare. For example, between 1 and 10^9 there are only 646 of these numbers, compared to 50,847,534 primes. The problem was finally settled in Alford, Granville, and Pomerance 1994, where it is shown that there are indeed infinitely many Carmichael numbers. A spin-off of their result has great relevance for primality testing and will be discussed in section 4.

3. Miller's test

In section 1 we saw that Fermat's theorem offers a way of checking that a given number is composite that does not require one to search for factors. However, this approach does not always work and, if one is very unlucky, it can fail quite miserably. Section 2 was an attempt to pin down what unluckiness means in this context. There we saw that Carmichael numbers behave as primes with respect to so many bases that it can be virtually impossible to spot their composite character with the test we have developed. But it is possible to improve this test so that it is not so easily fooled, even by Carmichael numbers. The new test was introduced by G. L. Miller in 1976.

Let $n > 0$ be an odd integer and choose an integer $1 < b < n - 1$, called the base, as before. Since n is odd, then $n - 1$ is even. The first step of *Miller's test* consists in finding $k \geq 1$ such that $n - 1 = 2^k q$, where q is an odd number. In other words, we must find the biggest power of 2 that divides $n - 1$, and also its co-factor q.

The test now proceeds with the computation of the residues of the following sequence of powers modulo n:

$$b^q, \ b^{2q}, \ldots, b^{2^{k-1}q}, b^{2^k q}.$$

Let's find out what properties this sequence has when n is a prime number. Thus, until further notice, we will be assuming that n *is prime*. Fermat's theorem tells us that

$$b^{2^k q} \equiv b^{n-1} \equiv 1 \pmod{n}.$$

Hence, if n is prime, the last residue in the sequence is always 1. Of course, an earlier element of the sequence could turn out to be 1. Suppose that j is the *smallest* exponent for which $b^{2^j q} \equiv 1 \pmod{n}$. If $j \geq 1$, we have

$$b^{2^j q} - 1 = (b^{2^{j-1} q} - 1)(b^{2^{j-1} q} + 1).$$

Since we are assuming that n is prime and that it divides $b^{2^j q} - 1$, it follows that either n divides $b^{2^{j-1} q} - 1$ or n divides $b^{2^{j-1} q} + 1$. But j is the smallest exponent for which $b^{2^j q} - 1$ is divisible by n. Therefore, n is not a factor of $b^{2^{j-1} q} - 1$. We conclude that $b^{2^{j-1} q} + 1$ must be divisible by n, so that $b^{2^{j-1} q} \equiv -1 \pmod{n}$.

The above argument shows that *if n is prime*, then one of the powers in the sequence

$$b^q, b^{2q}, \ldots b^{2^{k-1} q}$$

must be congruent to -1 modulo n. Well, not quite. The argument depends on j being greater than 0. If $j = 0$, then $b^q \equiv 1 \pmod{n}$. But we have no straightforward way of factoring $b^q - 1$ because q is odd. So, if *n is prime*, then one of two things can happen to the elements of the sequence of residues of powers modulo n: Either its first element is 1 or one of its elements is $n - 1$. If neither of these happens, then n must be composite.

The sequence of residues used in Miller's test is very easy to calculate because each residue (except the first) is the square of the one preceding it. Indeed, $b^{2^j q} = (b^{2^{j-1} q})^2$ for $j \geq 1$. As a consequence, if the sequence has a residue equal to $n - 1$, then all its successors will be congruent to 1 modulo n.

Once again, we have a test that, with luck, allows us to show that a given number is composite. However, Miller's test is far more efficient than the test of section 1. To understand why, note that if n is a pseudoprime to base b, then the sequence of powers must have an element congruent to 1 modulo n. But since n is not a prime, there is a good chance this power will not be preceded by an $n - 1$. In this case, Miller's test will spot that the number is composite. The algorithm for Miller's test is the following.

Miller's test

Input: an odd integer $n > 0$ and a base b, where $1 < b < n - 1$
Output: one of two messages: *"n is composite"* or *"inconclusive test"*

Step 1 Divide $n - 1$ by 2 as many times as necessary in order to find an odd co-factor. Thus we have found positive integers k and q, so that $n - 1 = 2^k q$ and q is odd.
Step 2 Begin by setting $i = 0$ and $r =$ residue of b^q modulo n.
Step 3 If $i = 0$ and $r = 1$, or if $i \geq 0$ and $r = n - 1$, the output is *"inconclusive test"*; otherwise go to step 4.
Step 4 Increase i by 1 and replace r by the residue of r^2 modulo n; go to step 5.
Step 5 If $i < k$, return to step 3; otherwise the output is *"n is composite"*.

In principle, when the output is *inconclusive*, two things can happen: Either n is prime or it is composite. Unfortunately, the second case actually occurs.

Let's look at a few examples, beginning with the the good news. We saw in section 1 that 341 is a pseudoprime to base 2, so this is a good target for Miller's test. First, $340 = 2^2 \cdot 85$. Now we must find the residues of the powers of 2 modulo 341 for the exponents 85 and 170:

$$2^{85} \equiv 32 \pmod{341}$$
$$2^{170} \equiv 32^2 \equiv 1 \pmod{341},$$

which means that the output is *composite*.

A more dramatic example is provided by the Carmichael number 561. Let's apply Miller's test to base 2. A simple calculation shows that $560 = 2^4 \cdot 35$. The sequence of residues of the powers of 2 modulo 561 is

Exponents	35	$2 \cdot 35$	$2^2 \cdot 35$	$2^3 \cdot 35$
Powers	263	166	67	1

Hence the output is *composite*. Although 561 is a Carmichael number, we spotted that it is composite using only the smallest possible base.

Now the bad news. Let's apply Miller's test to 25, using 7 as a base. Since $24 = 2^3 \cdot 3$, the sequence of residues is

Exponents	3	$2 \cdot 3$	$2^2 \cdot 3$
Powers	18	24	1

So the output of the test is inconclusive, even though 25 is composite. Of course 7 wasn't chosen as the base for nothing; if the base had been 2, the output would have been composite.

Let $n > 0$ be an odd integer and $1 < b < n - 1$. If n is composite, but has an inconclusive output for Miller's test to base b, then n is called a *strong pseudoprime* to base b. The example above shows that 25 is a strong pseudoprime to base 7. It is easy to see that if a number is a strong pseudoprime to base b, then it is a pseudoprime to this same base; see exercise 7.

As we have already said, 25 is not a strong pseudoprime to base 2. The smallest number that is a strong pseudoprime to base 2 is 2047. Moreover, there are only 1282 strong pseudoprimes to base 2 between 1 and 10^9. This gives a good measure of the efficiency of this test. Of course, we can apply Miller's test to several bases, thus increasing its efficiency to a very impressive standard. For example, the smallest strong pseudoprime to bases 2, 3, and 5 is 25,326,001.

Moreover, there are no "strong Carmichael numbers". This follows from a result of M. O. Rabin.

Rabin's theorem. *Let $n > 0$ be an odd integer. If, when Miller's test is applied to n, the output is "inconclusive test" for more than $n/4$ bases between 1 and $n - 1$, then n is prime.*

For details see Rabin 1980, theorem 1, and Knuth 1981, section 4.5.4, exercise 22. Needless to say, if n is large, it is not practicable to apply Miller's test to $n/4$ bases. Despite this fact, most practical primality tests are based upon Rabin's theorem, as we will see in the next section.

4. Primality testing and computer algebra

Many computer algebra systems have a simple command for testing whether a given number is prime. To the uninitiated the most surprising thing about this is that an answer is returned almost instantaneously, even when the given number is quite large. What happens is that most systems rely on Miller's test applied to a large enough set of bases to check whether or not a number is prime.

The rationale behind this use of Miller's test is provided by Rabin's theorem. Let n be an odd *composite* integer, and choose a base b, at random, between 1 and $n-1$. It follows from Rabin's theorem that the probability that Miller's test will have an inconclusive output when applied to n and b is less than or equal to

$$\frac{n/4}{n} = \frac{1}{4}.$$

So, it is plausible to assume that the probability of n being composite in this case is $1/4$. Now, if k bases are chosen, this probability should be $1/4^k$. Thus we can make the probability as small as we wish by choosing more and more bases.

These considerations lead to *Rabin's probabilistic primality test*. Since this will be a probabilistic test, we must decide how small we want the probability of making an error to be. Suppose that we decide that it should be smaller than ϵ, and let k be a positive integer such that $1/4^k < \epsilon$. Rabin's test consists in choosing k bases and applying Miller's test to every one of them. By the argument above, the probability that a composite number will have an inconclusive output for all these bases is smaller than or equal to $1/4^k$, and thus smaller than ϵ, which is what we wanted. How are we to choose these bases? Of course it is convenient to keep the bases small, otherwise the computations required by Miller's test will take too long to perform. A common choice is the first k positive primes.

Of course, if we want to be quite certain that the numbers tested will be prime, we have to choose ϵ very small. For example, let $\epsilon = 10^{-20}$. Since $1/4^{40}$ is of the order of magnitude of 10^{-24}, we might choose 40 bases to achieve a probability of error smaller than 10^{-20}. Let's assume that the bases are the first 40 primes. Unfortunately, there is a Carmichael number (of 397 digits) for which Miller's test has an inconclusive output when the chosen bases are the primes smaller than 300 (see Arnault 1995). Since there are 62 such primes, this number would have inconclusive output for all the bases we have chosen!

Let's see how Rabin's test is implemented in some well-known computer algebra systems. Of course, each system has its own strategy. For example, *Maple V.2*[1] tests primality in three stages. First, a search is made for prime factors smaller than 10^3. If no such factor is found, the system applies Miller's test to bases $2, 3, 5, 7$, and 11. Finally, it checks that the given number does not

---

[1] *Maple*[TM] is a product of Waterloo Maple Software, Inc.

belong to one of the following families:

$$(u+1)\left(k\frac{u}{2}+1\right) \quad \text{for} \quad 3 \le k \le 9, \quad \text{or} \quad (u+1)(ku+1) \quad \text{for} \quad 5 \le k \le 20$$

The reason for this last stage is that there are many known strong pseudoprimes in these families for the bases used as standard by the system (see Pomerance et al. 1980). However, the composite number

$$12,530,759,607,784,496,010,584,573,923$$

is identified as composite by *Maple V.2*. The smallest prime factor of this number is 286,472,803. The primality test has been modified in more recent versions of *Maple*, so that the number above is now correctly identified as composite.

Axiom 1.1[2] has a different strategy: It adjusts the number of bases to be used depending on the number that is going to be tested. It has been shown that the test used by *Axiom 1.1* always detects primality correctly for numbers smaller than 341,550,071,728,321 (see Jaeschke 1993). For numbers greater than this bound, the *Axiom 1.1* test uses the smallest 10 positive prime numbers as bases for Miller's test. Like *Maple*, the system also does some further checks for numbers that are considered to be especially troublesome. Once again the test is not perfect; it is defeated by a composite number of 56 digits.

At first, one might think that the strategy chosen for these systems would give a "perfect" test if only we could choose enough bases. But the truth is that this is not even theoretically possible. One of the consequences of the work of Alford, Granville, and Pomerance on Carmichael numbers is the following:

> Given any finite number of bases, there exist infinitely many Carmichael numbers that are strong pseudoprimes for all these bases.

Thus one should always beware of claiming that a number is prime on the basis of Miller's test applied to a fixed number of bases. One possible way out has been implemented in *Axiom 2.2*. The system now increases the number of bases to be chosen according to the size of the number that one wants to test. For a number of $2k$ decimal digits, the system chooses approximately k bases, thus increasing the accuracy of the test.

More details about how these systems test primality, and examples of the numbers that defeat each test can be found in Arnault 1995. In Chapter 10 we shall study tests that allow one to claim with certainty that a number is prime. As might be expected, these tests are neither as efficient nor as easy to use as Miller's test.

5. Exercises

1. Which of the following numbers are pseudoprimes to base 2: 645, 567, and 701? Which are pseudoprimes to base 3? Which are primes?

---

[2]*Axiom* is a registered trademark of NAG (Numerical Algorithms Group), Ltd.

2. Show that if n is a pseudoprime to bases a and ab, then n is also a pseudoprime to base b.

3. Let n be a positive integer, and set $p_1 = 6n + 1$, $p_2 = 12n + 1$, and $p_3 = 18n + 1$. Show that if p_1, p_2 and p_3 are primes, then the product $p_1p_2p_3$ is a Carmichael number. Show that these conditions are satisfied if $n = 1$, 6, and 35. Which Carmichael numbers are obtained for each of these values of n?

4. Factor 29,341 and show that it is a Carmichael number.

5. Let $p_1 < p_2$ be two odd primes. Write $n = p_1p_2$ and assume that $p_1 - 1$ and $p_2 - 1$ divide $n - 1$. Show that $n - 1 \equiv p_1 - 1 \pmod{p_2 - 1}$ and use this to obtain a contradiction. Conclude that a Carmichael number cannot have only two prime factors.

6. Which of the following integers are strong pseudoprimes to base 2: 645, 2047, and 2309? Which are strong pseudoprimes to base 3? Which are primes?

7. Show that if a positive odd integer n is a *strong* pseudoprime to base b, then it is a pseudoprime to this base.

8. Write a program to find all the pseudoprimes to bases 2 and 3 that are smaller than 10^6. Recall that n is a pseudoprime to bases 2 and 3 if it is *odd* and *composite* and satisfies

$$2^{n-1} \equiv 1 \pmod{n} \quad \text{and} \quad 3^{n-1} \equiv 1 \pmod{n}.$$

Hence the program will test the congruences only for integers that are odd and composite. One way to find these numbers is to run the sieve of Erathostenes for odd integers up to 10^6 and keep the composite numbers, instead of the primes. How many of the pseudoprimes you found are Carmichael numbers?

9. Write a program to find all Carmichael numbers with d prime factors, all of which are smaller than 10^3. The main problem is that, even for rather small values of d, some of the Carmichael numbers will be quite large. Thus it is necessary to use congruences to check that $p - 1$ divides $n - 1$, when p is a prime factor of n. Moreover, if the program has generated the number $n = p_1p_2 \ldots p_d$, then the residue of n modulo $p_i - 1$ is calculated multiplying the factors of n one at a time and reducing each product modulo $p_i - 1$. Since all the prime factors are smaller than 10^3, this keeps the numbers within the possibilities of the programming language. Use the program to find all Carmichael numbers with d factors smaller than 10^3 for $3 \leq d \leq 8$. It isn't necessary to multiply the factors; just list them for each number the program has obtained.

10. Write a program to find the smallest strong pseudoprime for each given base. The input will be an integer $b \geq 2$. The program should apply Miller's test (to base b) to all the composite odd integers until it finds one whose output is "inconclusive test". This will be the smallest strong pseudoprime to base b. Of course the search will be restricted to numbers smaller than the largest positive integer K supported by the programming language you have chosen. Thus the program can have one of two possible outputs: the smallest strong pseudoprime to base b or the message "there is no strong pseudoprime to base b smaller than K". To find the odd composite numbers smaller than K you can use the sieve of Erathostenes. Use the program to find the smallest strong pseudoprime to bases 2, 3, 5, and 7.

11. Write a program to find the pseudoprimes to base 2 that are of the form p^2, where p is a prime smaller than $5 \cdot 10^4$. The program will use the sieve of Erathostenes to find all positive primes $p \leq r$, and then test each prime to see which of them satisfy $2^{p^2} \equiv 2 \pmod{p^2}$. There are only two examples of square pseudoprimes to base 2 in the prescribed range.

7
Systems of Congruences

In this chapter we study a method for solving systems of linear congruences, the *Chinese remainder algorithm*. In the last section we will see how this algorithm is used to implement a scheme for sharing a secret key among several people.

1. Linear equations

Let's begin with the case of one linear equation,

$$(1.1) \qquad\qquad ax \equiv b \pmod{n}$$

where n is a positive integer. We saw in Chapter 4, section 7, that these equations are easy to solve when $\gcd(a,n) = 1$. For, by the invertibility theorem, $\gcd(a,n) = 1$ implies that $\overline{a}$ is invertible in $\mathbb{Z}_n$. Let $\overline{\alpha}$ be its inverse. Multiplying both sides of (1.1) by α, we obtain

$$\alpha(ax) \equiv \alpha b \pmod{n}.$$

Since $\alpha a \equiv 1 \pmod{n}$, it follows that

$$x \equiv \alpha b \pmod{n},$$

which is the solution of the equation. In particular, if n is prime and $a \not\equiv 0 \pmod{n}$, then (1.1) always has a solution.

Suppose now that $\overline{a}$ is *not* invertible in $\mathbb{Z}_n$, then $\gcd(a,n) \neq 1$. But if (1.1) has a solution, this means that there exist $x, y \in \mathbb{Z}$ such that

$$(1.2) \qquad\qquad ax - ny = b,$$

which is possible only if $\gcd(a,n)$ divides b. Thus, if (1.1) has a solution, then b is divisible by $\gcd(a,n)$. Of course if $\overline{a}$ has an inverse in $\mathbb{Z}_n$, this condition is satisfied because, in this case, $\gcd(a,n) = 1$.

Let's see if the converse of the conclusion we reached above also holds. Suppose that $d = \gcd(a,n)$ divides b. Thus $a = da'$, $b = db'$, and $n = dn'$, for some positive integers a', b' and n'. Hence, after canceling d, (1.2) becomes

$$a'x - n'y = b',$$

which is equivalent to the congruence $a'x \equiv b' \pmod{n'}$. Note that this is a congruence modulo n', which is a divisor of the original modulus n. Moreover, $\gcd(a',n') = 1$, so the new congruence must have a solution. Thus we have shown that if $\gcd(a,n)$ divides b, then (1.1) always has a solution.

Summing up, the congruence (1.1) has a solution if and only if $\gcd(a, n)$ divides b; see Chapter 1, exercise 7. Furthermore, this method for solving linear congruences is very easy to apply, since it uses only the extended Euclidean algorithm. However, when the solutions are obtained, we may be in for some surprises.

Let's solve the congruence $6x \equiv 4 \pmod 8$. Since $\gcd(6, 8) = 2 \neq 1$, it follows that $\overline{6}$ does not have an inverse in $\mathbb{Z}_8$. If the given congruence has a solution, then there must be integers x and y such that $6x - 8y = 4$. Dividing through by 2, we obtain $3x - 4y = 2$, which is equivalent to the congruence $3x \equiv 2 \pmod 4$. But $\overline{3}$ is its own inverse in $\mathbb{Z}_4$. Multiplying this last congruence by 3, we arrive at the solution

$$(1.3) \qquad\qquad\qquad x \equiv 2 \pmod 4.$$

This is not very satisfactory. After all, we began with a congruence modulo 8, so we would like to find the solutions modulo 8, not modulo 4 as in (1.3). This is easily remedied. It follows from (1.3) that if an integer x is a solution of $6x \equiv 4 \pmod 8$, then $x = 2 + 4k$ for some $k \in \mathbb{Z}$. If k is even, then $x \equiv 2 \pmod 8$ is one of the solutions. On the other hand, if k is odd, then $k = 2m + 1$ and $x = 6 + 8m$. Thus $x \equiv 6 \pmod 8$ is another solution. Moreover, since k is either even or odd, these are the only possibilities. Therefore, the equation $\overline{6} \cdot \overline{x} = \overline{4}$ has two distinct solutions in $\mathbb{Z}_8$, namely $\overline{2}$ and $\overline{6}$. Thus we have a *linear* equation with *two* solutions. As we saw in Chapter 5, section 4, this can only happen because, in this example, the modulus of the congruence is composite.

2. An astronomical example

In this section we describe a method for solving systems of linear congruences. This is a very old algorithm, which was used in antiquity to solve problems in astronomy. We begin with a problem of a modern bent that would have appealed to the ancient astronomers.

> Three satellites will cross the meridian of Leeds tonight: the first at 1 A.M., the second at 4 A.M., and the third at 8 A.M. Each satellite has a different period. The first takes 13 hours to complete one revolution around the Earth, the second takes 15 hours, and the third 19 hours. How many hours will pass (from midnight) until all three satellites cross the meridian of Leeds at the same time?

Let's see how the problem is translated into the language of congruences. Let x be the number of hours that have elapsed from 12 o'clock midnight tonight until the satellites cross the meridian of Leeds simultaneously. The first satellite crosses the meridian every 13 hours, starting at 1 A.M. Hence we must have $x = 1 + 13t$ for some integer t. In other words, $x \equiv 1 \pmod{13}$. The corresponding equations for the other two satellites are

$$x \equiv 4 \pmod{15} \quad \text{and} \quad x \equiv 8 \pmod{19}.$$

The three satellites will cross the meridian of Leeds simultaneously for the values of x that satisfy these three equations. Therefore, to answer the problem it is enough to solve the linear system of congruences:

(2.1)
$$x \equiv 1 \pmod{13}$$
$$x \equiv 4 \pmod{15}$$
$$x \equiv 8 \pmod{19}.$$

Note that we cannot add or subtract these equations, since the moduli are different. We will get over this problem by turning the congruences into equations involving integers.

Thus $x \equiv 1 \pmod{13}$ corresponds to $x = 1 + 13t$, which is an integer. Replacing x in the second equation by $1 + 13t$, we get

$$1 + 13t \equiv 4 \pmod{15} \quad \text{which is} \quad 13t \equiv 3 \pmod{15}.$$

But 13 is invertible modulo 15, and its inverse is 7. Multiplying $13t \equiv 3 \pmod{15}$ by 7 and replacing the numbers by their residues modulo 15, we have

$$t \equiv 6 \pmod{15}.$$

Hence t can be written in the form $t = 6 + 15u$ for some integer u. Therefore

$$x = 1 + 13t = 1 + 13(6 + 15u) = 79 + 195u.$$

Note that all the numbers of the form $79 + 195u$ are integer solutions of the first two congruences in (2.1). Finally, replace x by $79 + 195u$ in the last congruence of the system. We obtain

$$79 + 195u \equiv 8 \pmod{19} \quad \text{so that} \quad 5u \equiv 5 \pmod{19}.$$

Since 5 is invertible modulo 19, it can be canceled from the above equation, giving $u \equiv 1 \pmod{19}$. Rewriting this congruence as an equation in integers, we have $u = 1 + 19v$ for some integer v. Thus

$$x = 79 + 195u = 79 + 195(1 + 19v) = 274 + 3705v.$$

What can we conclude regarding the satellites? Recall that x is the number of hours that have elapsed from 12 midnight tonight until the satellites cross the meridian of Leeds simultaneously. Thus we have to find the smallest positive value of x that satisfies the three congruences; since $x = 274 + 3705v$, this is 274. Hence the three satellites will cross the meridian of Leeds simultaneously 274 hours after 12 midnight tonight. This corresponds to 11 days and 10 hours. But the general solution gives more information. Adding any multiple of 3705 to 274, we have another solution to the system. In other words, the satellites will cross the meridian simultaneously every 3705 hours after their first crossing; this corresponds to 154 days and 9 hours.

In the next section we will make a detailed analysis of the method used above for solving the linear system of congruences. Note that we solved this system of three congruences by solving two congruences at a time. Indeed, we first obtained the solution of the first two congruences, which is $x = 79 + 195u$.

This is equivalent to $x \equiv 79 \pmod{195}$. To find the solution of the three congruences, we then solved another system of two equations, namely

$$x \equiv 79 \pmod{195}$$
$$x \equiv 8 \pmod{19}.$$

In general, to solve a linear system with many congruences, we will have to solve several systems of two congruences. Thus, in the next section, we need only analyze in detail the algorithm used for solving systems of two congruences.

3. The Chinese remainder algorithm: Co-prime moduli

The Chinese remainder algorithm is so named because one of the first places where it can be found is *Master Sun's Mathematical Manual*, written between A.D. 287 and A.D. 473. In his book, Master Sun solved a numerical example, and then stated a rule for the solution of the same sort of problem. A more general analysis of the same problem, with several examples, can be found in the *Shu shu jiu zhang*, written by Qin Jiushaq in 1247. Similar problems can be found in the work of many other mathematicians, including the Indian Bhaskara (sixth century A.D.) and Nicomachus of Gerasa. For more details on the history of the theorem see Kangsheng 1988.

The Chinese remainder algorithm is merely a generalized version of the method used to solve the system of section 2. We analyze it in detail in this section.

Consider the system

(3.1)
$$x \equiv a \pmod{m}$$
$$x \equiv b \pmod{n}.$$

As in section 2, it follows from the first congruence that $x = a + my$, where y is an integer. Replacing x in the second congruence by $a + my$, we get $a + my \equiv b \pmod{n}$. In other words,

(3.2)
$$my \equiv (b - a) \pmod{n}.$$

But we know from section 1 that this equation has a solution if and only if the greatest common divisor of m and n divides $b - a$. To make sure that this condition holds, it is enough to assume that $\gcd(n, m) = 1$. Equivalently, $\overline{m}$ has an inverse in $\mathbb{Z}_n$; let's call it $\overline{\alpha}$.

It is now easy to solve (3.2). Multiplying both sides of the equation by α, we have $y \equiv \alpha(b - a) \pmod{n}$. Therefore, $y = \alpha(b - a) + nz$, where z is an integer. Since $x = a + my$, we obtain

$$x = a + m\alpha(b - a) + mnz.$$

But $\overline{\alpha m} = \overline{1}$ in $\mathbb{Z}_n$. Hence, there exists some integer β such that $1 - \alpha m = \beta n$. Thus

$$x = a(1 - m\alpha) + m\alpha b + mnz = a\beta n + m\alpha b + mnz.$$

Writing the solution of the system in this way has the advantage that α and β are easily computed. Indeed, $1 = \alpha m + \beta n$, so that α and β are found by applying the extended Euclidean algorithm to m and n. Summing up, if $\gcd(m, n) = 1$, then, for any given integer k, the numbers $a\beta n + b\alpha m + kmn$ are solutions of the system (3.1).

How many solutions does the above linear system have? Infinite, if we think of integer solutions. After all, for every choice of k we have a different solution, according to the formula we have obtained. But let's consider this point in more detail. Suppose that x and y are two integer solutions of (3.1). Then $x \equiv a$ (mod m) and $y \equiv a$ (mod m). Subtracting the second equation from the first, we conclude that $x - y \equiv 0$ (mod m). Equivalently, $x - y$ is divisible by m. Doing the same to the second congruence of the system, we have that $x - y$ is divisible by n. But $\gcd(m, n) = 1$, so, by the lemma of Chapter 2, section 4, $x - y$ is divisible by mn. Hence, if x and y are integer solutions of (3.1), then $x \equiv y$ (mod mn). Thus although the system has infinitely many integer solutions, they are all congruent modulo mn. In other words, the system has only one solution in $\mathbb{Z}_{mn}$. But we must not forget that this holds only because we are assuming that $\gcd(m, n) = 1$. Let's put all these facts together into a theorem.

Chinese remainder theorem. *Let m and n be co-prime positive integers. The system*

$$x \equiv a \quad (\text{mod } m)$$
$$x \equiv b \quad (\text{mod } n)$$

has one, and only one, solution in $\mathbb{Z}_{mn}$.

A good way to find out whether we have really understood this theorem is to consider its geometric interpretation. Suppose we have a table with mn entries. The columns of the table will be indexed by the elements of $\mathbb{Z}_m$, and the rows by the elements of $\mathbb{Z}_n$. If x is the entry in the intersection of the column indexed by $\overline{a} \in \mathbb{Z}_m$ with the row indexed by $\overline{b} \in \mathbb{Z}_n$, then

- $0 \leq x \leq mn - 1$,
- $x \equiv a$ (mod m), and
- $x \equiv b$ (mod n).

We will say that the entry x has *coordinates* $(\overline{a}, \overline{b})$ in this table. Since we are assuming that $0 \leq x \leq mn - 1$, we can think of these integers as representatives of the classes modulo mn. Thus x really stands for the class $\overline{x}$ of $\mathbb{Z}_{mn}$.

What does the Chinese remainder theorem tell us about this table? Assuming that $\gcd(m, n) = 1$, it follows from the theorem that every entry in the table corresponds to exactly one integer between 0 and $mn - 1$; that is, to one class of $\mathbb{Z}_{mn}$. Thus distinct entries have distinct coordinates, and vice versa. But do not forget that we are assuming that the moduli are co-prime. The table for $m = 4$ and $n = 5$ can be found on the next page.

Note that this table corresponds to the Cartesian product $\mathbb{Z}_4 \times \mathbb{Z}_5$. At first it may seem that, to find the entries of every cell in this table, it is necessary

	$\bar{0}$	$\bar{1}$	$\bar{2}$	$\bar{3}$
$\bar{0}$	$\bar{0}$	$\bar{5}$	$\overline{10}$	$\overline{15}$
$\bar{1}$	$\overline{16}$	$\bar{1}$	$\bar{6}$	$\overline{11}$
$\bar{2}$	$\overline{12}$	$\overline{17}$	$\bar{2}$	$\bar{7}$
$\bar{3}$	$\bar{8}$	$\overline{13}$	$\overline{18}$	$\bar{3}$
$\bar{4}$	$\bar{4}$	$\bar{9}$	$\overline{14}$	$\overline{19}$

to solve 20 linear systems of congruences. But the Chinese remainder theorem suggests that we ought to do it by "reverse engineering": Given an integer x between 0 and $mn - 1$, find its cell by computing its residue modulo m and its residue modulo n. Thus, in the example above, since 14 has residue 2 modulo 4, and residue 4 modulo 5, we conclude that it has coordinates $(\bar{2}, \bar{4})$.

But that's not the last word; we can do better. Indeed, it is possible to fill the whole table without calculating a single number. To understand how it's done, recall that we have a geometric interpretation for $\mathbb{Z}_4$: The four classes are represented by points equally spaced around a circumference. A similar representation holds for $\mathbb{Z}_5$.

Thus the table above is really like a map; it is a plane representation of a three-dimensional surface. To find this surface we proceed as follows: Since the classes of $\mathbb{Z}_4$ (the horizontal coordinates) are arranged around the circumference of the circle, we glue the right- and left-hand edges of the table. This produces a cylinder. But the classes of $\mathbb{Z}_5$ (the vertical coordinates) are also disposed on a circumference. Thus the top and bottom edges of the cylinder must also be glued. The resulting surface is called a *torus*, and it looks like a doughnut—of the sort that has a hole through the middle.

Let's go back to the task of finding the entries of each cell of the table. Since the entries 0, 1, 2, and 3 are smaller than both 4 and 5, they equal their own residues for both moduli. Thus we need not perform any calculations to find their coordinates. The table above, with these four entries in place, looks like this:

	$\bar{0}$	$\bar{1}$	$\bar{2}$	$\bar{3}$
$\bar{0}$	$\bar{0}$			
$\bar{1}$		$\bar{1}$		
$\bar{2}$			$\bar{2}$	
$\bar{3}$				$\bar{3}$
$\bar{4}$				

Note that as we move along the sequence of integers, we always go one column forward and one row down. The problem is that we have reached the right-hand edge of the table. If we had one more column, then 4 would find its place in it, one row down from where 3 is—that is, on the last row. But we do not have any more columns to the right, do we? That's where the geometric interpretation comes to our aid. Gluing the left-hand edge to the right-hand edge, we see that the first column on the left can be seen to the right of the last column. Back at the table, this means that we should "jump" from the last column to the first while, at the same time, moving one row down. Therefore 4 belongs at the intersection of the first column and the last row.

	$\bar{0}$	$\bar{1}$	$\bar{2}$	$\bar{3}$
$\bar{0}$	$\bar{0}$			
$\bar{1}$		$\bar{1}$		
$\bar{2}$			$\bar{2}$	
$\bar{3}$				$\bar{3}$
$\bar{4}$	$\bar{4}$			

However, we seem to have a new problem, because we have now reached the bottom edge. But, by the geometric interpretation, the bottom and top edges can also be glued. Thus we can go past the bottom edge into the first row. Only this time, we will be one column to the right of the previous cell. Doing this in the example, we get the following table:

	$\bar{0}$	$\bar{1}$	$\bar{2}$	$\bar{3}$
$\bar{0}$	$\bar{0}$	$\bar{5}$		
$\bar{1}$		$\bar{1}$		
$\bar{2}$			$\bar{2}$	
$\bar{3}$				$\bar{3}$
$\bar{4}$	$\bar{4}$			

Now we can carry on like this until all the entries are in place.

4. The Chinese remainder algorithm: General case

We have analyzed in great detail the solution of linear system of congruences when the moduli are co-prime because this is the only case to be used in later chapters. However, the Chinese remainder algorithm can also be used to solve systems whose moduli are not co-prime. In this case extra care is needed when solving the linear congruences that come up at every step of the algorithm. An example will suffice. Consider the system:

$$x \equiv 3 \quad (\text{mod } 12)$$
$$x \equiv 19 \quad (\text{mod } 8).$$

From the first equation we obtain $x = 3 + 12y$, for some integer y. Replacing x by $3 + 12y$ in the second equation, we have $12y \equiv 16 \ (\text{mod } 8)$. Since

$\gcd(12, 8) = 4$ divides 16, this latter congruence must have a solution. Indeed the congruence is equivalent to $12y - 8z = 16$ which, divided through by 4, gives $3y - 2z = 4$. Thus $3y \equiv 4 \pmod 2$. But $3 \equiv 1 \pmod 2$, so that $y \equiv 0 \pmod 2$. Therefore $y = 2k$, for some integer k. Finally, replacing y by $2k$ in $x = 3 + 12y$, we get $x = 3 + 24k$. It turns out that in this case there is only one solution modulo 24. However, $8 \cdot 12 = 96$. What is the relation between 24 and the two moduli 8 and 12? You will find the answer in exercise 5.

For a given pair of non-co-prime moduli, it is always possible to write a system that does not have any solutions. Thinking in terms of the geometric representation of section 3, this means that if the moduli are not co-prime then there will always be blank cells left in the table.

Once more, it is not necessary to do any calculations in order to fill in the table. All we have to do is fill the cells with the integers $0, 1, \ldots$ always moving one column forward and one row down at every step; not forgetting the "jumps": right to left and bottom to top. If we do this when the moduli are not co-prime, we get back to the entry with coordinates $(\overline{0}, \overline{0})$ before we reach $mn - 1$. This explains why some entries will be left empty. When $m = 4$ and $n = 6$, the table is as follows:

	$\overline{0}$	$\overline{1}$	$\overline{2}$	$\overline{3}$
$\overline{0}$	$\overline{0}$		$\overline{6}$	
$\overline{1}$		$\overline{1}$		$\overline{7}$
$\overline{2}$	$\overline{8}$		$\overline{2}$	
$\overline{3}$		$\overline{9}$		$\overline{3}$
$\overline{4}$	$\overline{4}$		$\overline{10}$	
$\overline{5}$		$\overline{5}$		$\overline{11}$

5. Powers, again

There is a version of the Chinese remainder theorem for more than two equations. We will state it, but leave the proof to you, since it is just another application of the Chinese remainder algorithm. First, a definition: The positive integers $n_1, \ldots, n_k$ are said to be *pairwise co-prime* if $\gcd(n_i, n_j) = 1$, whenever the indices i and j are different. For example, three moduli, n_1, n_2, n_3, are pairwise co-prime when $\gcd(n_1, n_2) = \gcd(n_1, n_3) = \gcd(n_2, n_3) = 1$.

Chinese remainder theorem. *Let $n_1, \ldots, n_k$ be pairwise co-prime positive integers. The system*

$$x \equiv a_1 \pmod{n_1}$$
$$x \equiv a_2 \pmod{n_2}$$
$$\vdots$$
$$x \equiv a_k \pmod{n_k}$$

has one, and only one, solution in $\mathbb{Z}_{n_1 \ldots n_k}$.

We can use the Chinese remainder theorem to simplify the computation of residues of powers modulo n when the complete factorization of n is known. We will also assume that each prime factor has multiplicity 1 in the factorization of n, because this is the case for which the method is most effective.

Suppose that n has been factored and that $n = p_1 \ldots p_k$, where $0 < p_1 < \cdots < p_k$ are prime numbers. Given positive integers a and m, we first compute the residue of a^m modulo each prime factor of n. If the prime factors are not too large, the computation is very fast. This is true even when m and a are large, because Fermat's theorem comes to our aid. Suppose that we have carried out these calculations, and that

$$a^m \equiv r_1 \pmod{p_1} \quad \text{and} \quad 0 \le r_1 \le p_1$$
$$a^m \equiv r_2 \pmod{p_2} \quad \text{and} \quad 0 \le r_2 \le p_2$$
$$\vdots$$
$$a^m \equiv r_k \pmod{p_k} \quad \text{and} \quad 0 \le r_k \le p_k.$$

Thus, to find the residue of a^m modulo n, we have only to solve the system

$$x \equiv r_1 \pmod{p_1}$$
$$x \equiv r_2 \pmod{p_2}$$
$$\vdots$$
$$x \equiv r_k \pmod{p_k}.$$

Note that since the moduli of this system are distinct primes, they are necessarily pairwise co-prime. Hence, by the Chinese remainder theorem the system always has a solution, say, $0 \le r \le n - 1$. Moreover, any two such solutions are congruent modulo $p_1 \cdots p_k = n$. Since a^m is also a solution of the system, we must have $a^m \equiv r \pmod{n}$. Thus r is the residue of a^m modulo n.

Here's an example. Suppose that we want to find the residue of 2^{6754} modulo 1155. Factoring 1155, we find that it is equal to $3 \cdot 5 \cdot 7 \cdot 11$. Using Fermat's theorem for each of these primes, we obtain

$$2^{6754} \equiv 1 \pmod{3}$$
$$2^{6754} \equiv 4 \pmod{5}$$
$$2^{6754} \equiv 2 \pmod{7}$$
$$2^{6754} \equiv 5 \pmod{11}.$$

Thus, we must solve the system

$$x \equiv 1 \pmod{3}$$
$$x \equiv 4 \pmod{5}$$
$$x \equiv 2 \pmod{7}$$
$$x \equiv 5 \pmod{11}$$

by the Chinese remainder algorithm. Since $x = 1 + 3y$, the second congruence becomes

$$1 + 3y \equiv 4 \pmod{5}, \quad \text{so that} \quad y \equiv 1 \pmod{5},$$

because 3 is invertible modulo 5 and can be canceled from both sides of the equation. Thus $x = 4 + 15z$. Replacing x by $4 + 15z$ in the third congruence and solving it, we obtain $x = 79 + 105t$. Finally, solving the fourth congruence for t, we have $t \equiv 6 \pmod{11}$. Hence $x = 709 + 1155u$, and 709 is the residue of 2^{6754} modulo 1155.

6. On sharing secrets

Benjamin Franklin once said, "Three may keep a secret, if two of them are dead"; in this section we study a secure system for sharing a secret among living beings, based on the Chinese remainder theorem. Imagine the following scenario: A bank vault must be opened every day, and the bank employs five senior tellers who have access to it. But, for security reasons, management would prefer a system that required at least two of the five senior tellers to be present for access to the vault to be granted. Of course, the problem is that access should be granted when *any* two of the five senior tellers are in the bank, but not otherwise.

Let's consider the same problem in greater generality. In order to gain access to the bank vault it is necessary to know a key, which we can assume to be a positive integer s. We want to share this key among the n senior tellers, so that each will know something about s. Let's call this partial information a *piece* of the key. Moreover, access to the vault should *not* be possible unless at least k senior tellers are in the bank, where $k \geq 2$ is a positive integer smaller than n. We achieve this by sharing the key so that

- it is *easy* to find s if k or more pieces are known, and
- it is *difficult* to find s if less than k pieces are known.

The pieces of the key that each person receives are really elements of a set $\mathbb{S}$ of n *ordered pairs* of positive integers. To construct $\mathbb{S}$ with the required properties we first choose a set $\mathcal{L}$ of n pairwise co-prime positive integers. Let N be the product of the k *smallest* numbers of $\mathcal{L}$, and let M be the product of the $k - 1$ *largest* numbers of $\mathcal{L}$. We say that $\mathcal{L}$ has *threshold k* if $M < N$. It follows from this condition that the product of any k (or more) elements of $\mathcal{L}$ is always *bigger* than N, and that the product of less than k of its elements is always *smaller* than M.

Suppose that the key s has been chosen so that $M < s < N$, and let $\mathbb{S}$ be the set of numbers of the form (m, s_m), where $m \in \mathcal{L}$ and s_m is the residue of s modulo m. These pairs are the *pieces of the key* the senior tellers will receive. The fact that we have a set $\mathcal{L}$ with threshold $k \geq 2$ implies that $s > m$ for every $m \in \mathcal{L}$. In particular, $s_m < s$ for every $m \in \mathcal{L}$.

What happens if k or more of the senior tellers are in the bank? In this case, t of the n pairs in $\mathbb{S}$ are known for some $t \geq k$. If the pairs are

$(m_1, s_1), \ldots, (m_t, s_t)$, then we consider the linear system

$$x \equiv s_1 \quad (\text{mod } m_1)$$
$$x \equiv s_2 \quad (\text{mod } m_2)$$

(6.1)

$$\vdots$$

$$x \equiv s_t \quad (\text{mod } m_t).$$

The elements of $\mathcal{L}$ are pairwise co-prime. So, by the Chinese remainder theorem, the system has a solution $0 \leq x_0 < m_1 \ldots m_t$. But is x_0 equal to s? This is the point in the argument where we need to know that $\mathcal{L}$ has threshold k. Since $t \geq k$, it follows that

$$m_1 \ldots m_t \geq N > s.$$

But s is also a solution of (6.1), so by the Chinese remainder theorem,

$$x_0 \equiv s \quad (\text{mod } m_1 \ldots m_t).$$

Since s and x_0 are positive integers smaller than $m_1 \ldots m_t$, we have $s = x_0$.

Suppose now that fewer than k senior tellers are in the bank. Then, although t is now smaller than k, we can still solve (6.1). Let x_0 be its smallest non-negative solution; then $0 \leq x_0 < m_1 \ldots m_t$. But the product of fewer than k elements of $\mathcal{L}$ is always smaller than s; hence $x_0 < M < s$. Therefore, it is not enough to solve the system in order to find s. However, x_0 and s are both solutions of (6.1), so that

$$s = x_0 + y \cdot (m_1 \ldots m_t),$$

where y is a positive integer. Since

$$N > s > M > x_0,$$

it follows that

$$\frac{M - x_0}{m_1 \ldots m_t} \leq y = \frac{s - x_0}{m_1 \ldots m_t} \leq \frac{N - x_0}{m_1 \ldots m_t}.$$

Using $t < k$, we conclude that it will be necessary to search for y among at least

$$d = \left\lceil \frac{N - M}{M} \right\rceil$$

integers. Choosing the moduli so that d is very large makes this search utterly impractical.

We have been left with a problem, though: Can $\mathcal{L}$ be chosen so that all these conditions are satisfied? The answer is yes, but it uses results on the distribution of primes that we have not dealt with in this book. For a good discussion of this point see Kranakis 1986, Chapter 1, section 5.

Let's review the construction. The data we need consist of the number n of senior tellers who have access to the bank vault, and the smallest number of these who must be in the bank for the system to grant them access to the vault. The first number determines the size of $\mathcal{L}$; the second, its threshold k. Next, we must choose a set $\mathcal{L}$ with n elements and with threshold k—this is the part of the construction we did not discuss in detail—and compute the numbers M and N defined above. Recall that $\mathcal{L}$ must be chosen so that the number d above is very large; otherwise the key can be found by a simple search. The key s is an integer chosen randomly in the interval between M and N. Now we can compute the elements of $\mathbb{S}$ and share them among the staff. Of course the security of this scheme depends on the fact that the bigger k is, the less likely it is that one can find k dishonest senior tellers in the same bank. If they're all dishonest, we are lost; no system is 100 percent secure.

Here's an example. Suppose that there are five senior tellers in the bank, and that at least two must be present for the security system to let them open the vault. Thus $\mathcal{L}$ must be a set with 5 elements, and its threshold must be 2. Choosing the elements of $\mathcal{L}$ among the smaller primes, we have

$$\mathcal{L} = \{11, 13, 17, 19, 23\}.$$

The product of the two smallest primes in this set is $N = 11 \cdot 13 = 143$. On the other hand, since $k = 2$, the product of the $k - 1$ largest primes in $\mathcal{L}$ is actually equal to the largest element of $\mathcal{L}$. Thus $M = 23$, and $\mathcal{L}$ has threshold 2. Now s can be any integer between 23 and 143, say, $s = 30$. Then

$$\mathbb{S} = \{(11, 19), (13, 17), (17, 13), (19, 11), (23, 7)\}.$$

Finally, what happens if the senior tellers with pieces $(17, 13)$ and $(23, 7)$ are in the bank? They key in their pieces and the security system solves the system

$$x \equiv 13 \pmod{17}$$
$$x \equiv 7 \pmod{23},$$

and finds that its smallest positive solution is 30. This is the correct key, so access to the vault is granted.

7. Exercises

1. Solve the following system of congruences, which appears in the *Yih-hing* (A.D. 717):

$$x \equiv 1 \pmod{2}$$
$$x \equiv 2 \pmod{5}$$
$$x \equiv 5 \pmod{12}$$

2. A problem from *Master Sun's Mathematical Manual*: There are certain things whose number is unknown. If the number is repeatedly divided by 3, the remainder is 2; divided by 5, the remainder is 3; and divided by 7, the remainder is 2. What will the number be?

3. A problem from the *Aryabhatiya*, an Indian arithmetical tract of the sixth century: Find the smallest positive number that if divided by 8 is known to leave 5; that if divided by 9, leaves a remainder 4; and that if divided by 7, leaves a remainder 1.

4. In ancient Indian astronomy, a *Kalpa* was a period of 4320 million years at the beginning and end of which all the fundamental astronomical numbers of the planets were assumed to be zero. Suppose that at a certain time since the beginning of the *Kalpa*, the sun, moon, etc., have traveled for the following number of days after completing their full revolutions:

Sun	Moon	Mars	Mercury	Jupiter	Saturn
1000	41	315	1000	1000	1000

Given that the sun completes three revolutions in 1096 days, the moon five revolutions in 137 days, Mars one revolution in 185 days, Mercury thirteen revolutions in 1096 days, Jupiter three revolutions in 10,960 days, and Saturn one revolution in 10,960 days, find the number of days since the *Kalpa*.

5. Show that the system

$$x \equiv a \pmod{m}$$
$$x \equiv b \pmod{n}$$

cannot have more than one solution modulo the least common multiple of m and n. Note that we are *not* assuming that m and n are co-prime.

6. Use the Chinese remainder theorem to compute the remainder of the division of $2^{45,632}$ and of $3^{54,632}$ by 12,155.

7. Solve the equation $x^2 + 42x + 21 \equiv 0 \pmod{105}$.
Hint: Factor 105, solve the equation modulo each prime factor, and then use the Chinese remainder algorithm.

8. Find the longest sequence of *consecutive primes* with threshold 3 and first element 11 Do the same for threshold 4.

9. Let p and q be distinct primes and $n = pq$. Suppose that we know the solutions of the equations $x^2 \equiv a \pmod{p}$ and $x^2 \equiv a \pmod{q}$. Show how the Chinese remainder algorithm can be used to find a solution of $x^2 \equiv a \pmod{n}$. Compare your solution with the method of section 5 and with exercise 7.

10. Let p and q be distinct primes and $n = pq$. Suppose that both primes leave remainder 3 in the division by 4. Write a program that, having p, q and a as input, computes the solutions of $x^2 \equiv a \pmod{n}$. The restriction on the primes makes it easier to solve the equations $x^2 \equiv a \pmod{p}$ and $x^2 \equiv a \pmod{q}$; see Chapter 5, exercise 17. This is the third exercise in a series that ends with exercise 8 of Chapter 11.

8

Groups

In this chapter we introduce groups and subgroups, and prove Lagrange's theorem. Groups are one of the "taxonomic classes" we use to classify mathematical structures that have common characteristics. As with any category in a classification scheme, we will only be able to understand what a group is if we are familiar with many particular examples. The examples discussed in this chapter include the groups of symmetries of polygons, and the group of invertible integers modulo n. This last example is the key to the applications of groups to number theory in Chapters 9 and 10.

1. Definitions and examples

A *group* has two basic ingredients: a *set* and an *operation* defined in this set. Let's denote the set by G and the operation by $\star$. By *operation* we understand a rule for combining any two elements a and b of G to get another element of G, which we denote by $a \star b$.

It is often the case in mathematics that a set comes equipped with an operation. Familiar examples include the natural numbers with addition, the integers with addition, the rational numbers with multiplication, and vectors in 3-space with the vector product.

However, not every set with an operation is a group. A set G with an operation, denoted by $\star$, is a *group* if this operation satisfies the following properties:

- **Associativity:** Given any elements $a, b, c \in G$, we have
$$a \star (b \star c) = (a \star b) \star c.$$

- **Identity element:** There exists an element $e \in G$ such that, for all $a \in G$,
$$a \star e = e \star a = a.$$

- **Inverse:** Given any $a \in G$, there exists an element $a' \in G$, called the *inverse* of a, such that
$$a \star a' = a' \star a = e.$$

The reason why groups are defined like this is that sets with operations that satisfy these properties are ubiquitous and have some very nice properties. Thus groups are so defined, not by divine dispensation, but for purely pragmatical reasons.

Note that we are *not* requiring the operation to be commutative. In other words, it need not be true that $a \star b = b \star a$ for every $a, b \in G$. Once again, the reason for this choice is purely a matter of convenience: There are many interesting groups whose operations are not commutative. When the operation of a group is commutative, we say that the group is *abelian*.

There are many sets with an operation that are not groups. For example, the addition of natural numbers is associative and has zero as its identity element. However, the only natural number that has an inverse is 0, because negative numbers are not in the set $\mathbb{N}$.

An even more dramatic example is the set of vectors in 3-space with the vector product. The set is not a group because the vector product is not an associative operation. By the way, the inner product of vectors is not an operation at all, in our sense of the word, because it combines two vectors to produce a number, not a vector.

On the other hand, groups abound among the most familiar sets with operations. For example, $\mathbb{Z}$, $\mathbb{Q}$, $\mathbb{R}$, and $\mathbb{C}$ are groups with respect to addition. Since the only integers with multiplicative inverses are ± 1, the set $\mathbb{Z}$ is not a group under multiplication; nor are $\mathbb{Q}$, $\mathbb{R}$, and $\mathbb{C}$, because division by zero is not possible. However, if 0 is removed, these last three sets become groups. Thus $\mathbb{Q} \setminus \{0\}$, $\mathbb{R} \setminus \{0\}$, and $\mathbb{C} \setminus \{0\}$ are groups with respect to multiplication.

For any positive integer n, the set $\mathbb{Z}_n$ is a group under the operation of addition. The set of square $n \times n$ matrices with real coefficients is a group under the addition of matrices, while the invertible matrices (those whose determinant is non-zero) form a group under the multiplication of matrices. Note that the latter is *not* an abelian group, because multiplication of matrices is not commutative.

The number of elements of a group is its *order*. The groups mentioned above are all infinite, with the exception of $\mathbb{Z}_n$, which has order n. Another well-known finite group is $\{-1, 1\}$ under the multiplication of integers; it has order 2. Notice that we have been saying "group" when we really mean the underlying set. That's the common usage, and we will go along with it whenever possible. We will study more interesting examples of finite groups in the next two sections.

One final comment on terminology: Many of the results in this chapter will refer to a "general" group. Thus, to avoid confusion, it is convenient to go on using a neutral symbol like $\star$ for the operation of a general group. However, we will still say "multiply" and "multiplying" even though the operation is $\star$ and not multiplication. The reason is that the neologisms "star" and "starring" are far too ugly to contemplate.

2. Symmetries

One of the most important applications of groups is to the study of symmetries. One could even say that groups are the translation of the concept of symmetry in the language of mathematics. Thus, not surprisingly, they play a key role in many disciplines where symmetries are fundamental, like geometry, crystallography, and physics.

Having said that, we must face the fact that our general concept of symmetry is very elusive. In geometry a symmetry of a figure is a transformation that, when applied to the points of the figure, does not alter its appearance. A better way of putting this might be as follows. Imagine that you are looking at a geometric figure, say, a polygon. Now close your eyes while someone applies a transformation to the figure. If, when you open your eyes, you are unable to tell whether or not anything has been done to the figure, then that transformation is a symmetry. This may still seem very vague, but it is good enough to handle the simple examples of this chapter. For a thorough discussion of symmetry in science and art see Weyl 1982.

Let's try to find all the symmetries of an equilateral triangle. First of all, we have three counterclockwise rotations, of 120°, 240°, and 360°. The last one coincides with the rotation of 0°. There are also three reflections. Each of these has as its axis (or mirror) one of the lines that bisects an angle of the triangle. It is clear that these six transformations satisfy the criterion of the previous paragraph, so they are symmetries of the equilateral triangle. Moreover, it can be shown that these are all the symmetries of an equilateral triangle—we will have more to say about this at the end of the section.

We have the set, but the operation is missing. If we think of the symmetries as transformations of the set of points that form the triangle, then the operation is the composition of symmetries. Since the composition of maps is always associative, the first property of the operation of a group is clearly satisfied. The role of the identity element is played by the rotation of 0°, which is really the transformation that consists of not doing anything to the triangle at all.

What about the inverses? The inverse of a rotation of 120° is the rotation of 240°, and vice versa. The reason is that $120 + 240 = 360$, and a rotation of 360° is essentially the same as a rotation of 0°. Each reflection is clearly its own inverse. Hence all the symmetries described above have an inverse, and the set of symmetries of an equilateral triangle, with the composition of symmetries, is a group of order 6 usually denoted by D_3.

Let's attach numbers to the vertices of the equilateral triangle: 1 and 2 will be the vertices of the base, and 3 the top vertex, as shown in the figure.

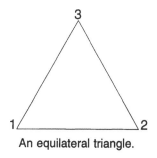

An equilateral triangle.

We can now describe the symmetries of the triangle as permutations of the vertices. For example, the rotation of $120°$ takes each vertex to the place of its adjacent vertex in the counterclockwise direction. There is a very practical notation for describing this permutation:

$$\begin{pmatrix} 1 & 2 & 3 \\ 2 & 3 & 1 \end{pmatrix}.$$

This is the transformation that, when applied to the triangle, moves vertex 1 to the place where vertex 2 originally was, vertex 2 to the place where 3 originally was, and 3 to the place where 1 originally was. On the top row we always write 1, 2, and 3 in order; on the bottom row we write the place where each vertex ended up after the transformation was applied to the triangle. Note that a place is named after the vertex that occupied it before we applied the transformation. Here is another example. The reflection about the line that bisects vertex 3 is

$$\begin{pmatrix} 1 & 2 & 3 \\ 2 & 1 & 3 \end{pmatrix}.$$

If ρ is a rotation of $120°$, then

$$\rho^2 = \rho\rho$$

is a rotation of $240°$, and $\rho^3 = e$ is the identity element—a rotation of $360°$. If σ is any one of the reflections, then $\sigma^2 = e$. We wish to identify the symmetry that corresponds to $\sigma\rho$. Note that $\sigma\rho$ cannot be equal to ρ^2. Indeed, "multiplying" $\sigma\rho = \rho^2$ on the right-hand side by ρ^2 and using $\rho^3 = e$, we have $\sigma = \rho$, which is a contradiction. Similarly, we can show that $\sigma\rho \neq e$ and that $\sigma\rho \neq \rho$. Hence $\sigma\rho$ cannot be a rotation, and $\sigma\rho$ must be a reflection. However, $\sigma\rho \neq \sigma$, since $\sigma\rho = \sigma$ implies that $\rho = e$, another contradiction. Therefore, $\sigma\rho$ must be a reflection different from σ.

Let's denote by σ_3 the reflection that does not move vertex 3, which we described above. Then ρ moves vertex 1 to the position of vertex 2, while σ_3 moves vertex 2 to the position of 1. Thus $\sigma_3\rho(1) = 1$; in other words, vertex 1 stays put under $\sigma_3\rho$. We will denote the reflection that does not move vertex 1 by σ_1. Therefore, $\sigma_3\rho = \sigma_1$.

We could also have computed $\sigma_3\rho$ using the notation introduced above; thus

$$\sigma_3 = \begin{pmatrix} 1 & 2 & 3 \\ 2 & 1 & 3 \end{pmatrix} \quad \text{and} \quad \rho = \begin{pmatrix} 1 & 2 & 3 \\ 2 & 3 & 1 \end{pmatrix}.$$

Before we carry out the calculation, note that $\sigma_3\rho(1)$ means that ρ is applied to 1 *before* σ_3, so that

$$1 \xrightarrow{\rho} 2 \xrightarrow{\sigma_3} 1.$$

Hence

$$\sigma_3\rho = \begin{pmatrix} 1 & 2 & 3 \\ 2 & 1 & 3 \end{pmatrix}\begin{pmatrix} 1 & 2 & 3 \\ 2 & 3 & 1 \end{pmatrix} = \begin{pmatrix} 1 & 2 & 3 \\ 1 & 3 & 2 \end{pmatrix} = \sigma_1.$$

Using only the basic properties of a group, we can compute several other relations between the elements of D_3 taking $\sigma_3\rho = \sigma_1$ as a starting point. For example, let G be a group, and let $\star$ be its operation. If $x, y \in G$, then the inverse of $x \star y$ is $y' \star x'$. To check that this is the case, it is enough to multiply these two elements; so,

$$(x \star y) \star (y' \star x') = x \star (y \star y') \star x' = x \star e \star x' = x \star x' = e.$$

Using this simple fact and continuing to use a dash to denote the inverse of an element, we have

$$(\sigma_3 \rho)' = \rho^2 \sigma_3.$$

But we have seen that $\sigma_3 \rho = \sigma_1$. Since $\sigma_1^2 = e$, we conclude that $\rho^2 \sigma_3 = \sigma_1$. Therefore,

$$\sigma_3 \rho = \sigma_1 = \rho^2 \sigma_3 \neq \rho \sigma_3.$$

In particular, D_3 is not an abelian group.

There are many other relations that follow from $\sigma_3 \rho = \sigma_1$. Multiplying it on the left by σ_3 and using $\sigma_3^2 = e$, we have $\rho = \sigma_3 \sigma_1$; while multiplying it on the right by ρ^2 we have $\sigma_3 = \sigma_1 \rho^2$. Note that since the operation of D_3 is not commutative, we must specify which of the two sides of the equation is to be multiplied by the given element.

Carrying these calculations far enough, we can fill in the multiplication table of D_3. In general, the *multiplication table* of a finite group is a table whose rows and columns are indexed by the elements of the group. If the group operation is $\star$, then the entry of the cell in the intersection of the row indexed by x with the column indexed by y is $x \star y$. The multiplication table of the group D_3 is as follows:

	e	ρ	ρ^2	σ_1	σ_2	σ_3
e	e	ρ	ρ^2	σ_1	σ_2	σ_3
ρ	ρ	ρ^2	e	σ_3	σ_1	σ_2
ρ^2	ρ^2	e	ρ	σ_2	σ_3	σ_1
σ_1	σ_1	σ_2	σ_3	e	ρ	ρ^2
σ_2	σ_2	σ_3	σ_1	ρ^2	e	ρ
σ_3	σ_3	σ_1	σ_2	ρ	ρ^2	e

Note that there are no repeated elements in either row or column of this table. This is a general fact, which is true for the multiplication table of any group. To prove it, suppose that we have a group G whose operation is $\star$. The entries of the row indexed by $a \in G$ are of the form $a \star x$ for some $x \in G$. Now, if there exist $x, y \in G$ such that $a \star x = a \star y$, then

$$x = a' \star (a \star x) = a' \star (a \star y) = y,$$

where a' is the inverse of a in G. Thus, the entries $a \star x$ and $a \star y$ are equal only if they belong to the same column. In particular, the entries of a given row of the multiplication table of a group G must be distinct. A similar argument proves the corresponding result for columns.

In general, the symmetry group of a regular polygon of n sides, denoted by D_n, has order $2n$ and is generated by the rotation ρ of $360/n$ degrees and by any one of the reflections. If σ is a reflection, then

$$\sigma\rho = \rho^{n-1}\sigma.$$

This group is called the dihedral group of order $2n$. To show that this description corresponds to the whole group of symmetries of a regular polygon it is necessary to use linear algebra; you will find the details in Artin 1991, Chapter 5.

3. Interlude

The theory of groups is a relatively recent branch of mathematics that grew out of the theory of polynomial equations. Quadratic equations were routinely solved by the Babylonians more than a thousand years before Christ. The Greeks, more interested in geometry, did not contribute much to this subject. The interest in equations was revived by the Arabs, who looked for ways of solving polynomial equations of the third degree, or cubics. The real breakthrough, however, came only in Renaissance Italy.

The history of the discovery of the formulae for solving polynomial equations of degrees 3 and 4 is rife with intrigue and betrayal. It all began with Scipione del Ferro, a professor at the University of Bologna, one of the oldest of the Medieval universities. It is not known when del Ferro discovered the solution of the cubic, but before his death around 1526, he explained the method to his student, Antonio Maria Fior.

This was a time when competitions among the learned were common, and Fior decided to challenge Niccolò Tartaglia ("the stutterer"), a professor of mathematics at Venice. The contest would consist of 30 questions, and the loser would pay for 30 banquets. Close to the day when the allotted time was to expire, Tartaglia had an inspiration: He discovered on his own the method for finding the roots of cubic equations, and solved within hours all the proposed problems. Fior did not fare so well, and was unable to solve most of the problems posed by his adversary. Thus Tartaglia was declared the winner and, in a gentlemanly fashion, renounced the 30 banquets—the honor of winning was enough for him.

Tartaglia's triumph led to an invitation to visit Gerolamo Cardano, who was famous as a doctor, mathematician, and astrologer. He told Cardano the details of his solution of the cubic, but made Cardano swear that he would keep it secret. Cardano extended the method and finally published it in his *Ars magna* in 1545. Besides the solution of the cubic, the book contained a method for reducing the solution of a biquadratic equation to that of a cubic. This last result was obtained by Lodovico Ferrari, Cardano's friend and secretary.

Needless to say, Tartaglia was incensed. He accused Cardano of perjury, and even published the whole story, including the complete text of the oath. What had happened, however, was that Cardano had in the meantime found out that del Ferro had discovered the same results before Tartaglia. This, in his view, left him free to publish the results. In the book he states clearly that the solution of the cubic had first been found by del Ferro, and had later been rediscovered by Tartaglia.

For the next 300 years mathematicians searched in vain for similar methods for solving equations of degree greater than 4. What they wanted was a way to find the roots by a sequence of operations applied to the coefficients of the polynomial equation. However, only the operations of addition, subtraction, multiplication, division, and extraction of roots were to be allowed. This is known in the trade as *solving equations by radicals*.

The truth is that these restrictions make the problem insoluble. The first complete proof of this fact was given by the Norwegian mathematician N. H. Abel in 1824. Abel contributed to many areas of mathematics, notably analysis and algebraic geometry, so that now we talk of abelian groups, abelian functions, and Abel's theorem on the convergence of series. His achievement is all the more surprising because he died of consumption before his twenty-seventh birthday. So great is his fame that a statue in his honor has been raised in the Royal Park in the center of Oslo.

The complete answer to the problem of solving polynomial equations by radicals was found by Abel's contemporary, E. Galois. He showed that to each polynomial equation there corresponds a finite group, which completely determines whether or not the equation can be solved by radicals. Since the group is finite, this can, at least in principle, be turned into an algorithm.

Galois's life was even more tragic than Abel's. His father committed suicide for political reasons; his work was considered incomprehensible by the members of the Académie des Sciences de Paris; and he did not manage to pass the entrance examination of the École Polytechnique. In fact, he was so furious at the stupidity of the Polytechnique examiners that he threw the blackboard eraser at one of them! Since he needed financial assistance, he decided to enter the École Préparatoire, a teacher training college. Galois was also ardently republican at a time when France was a monarchy, and his political activities finally led to his expulsion from the École Préparatoire.

He was in prison for his political activities when, during a cholera epidemic in 1832, he was transferred to a hospital. There he fell in love with a girl, but little is known of the affair. The fact is that soon thereafter he was challenged to a duel. The reasons why the duel was fought are not clear. A recent reconstruction suggests that, depressed by the failure of his affair and by lack of recognition for his work, he may have offered to die for the republican cause. The idea was to pretend that he had been killed by loyalists, and use that as a reason for starting an uprising during his funeral. Even there, tragedy struck. During the funeral the leaders heard of the death of General Lamarque. A larger crowd would gather for the funeral of the famous general, and it was quickly decided that it would be better to delay the uprising until then. Thus nothing happened at Galois's funeral.

E. Galois (1811–1832).

The night before the duel, aware that death was near, Galois wrote a last-minute letter to his friend, Auguste Chevalier. In this letter, after reviewing his discoveries, he concluded:

> Ask Gauss or Jacobi publicly to give their opinion, not as to the truth, but as to the importance of these theorems. Later on there will be, I hope, people who will find some profit in sorting out this mess.

Mortally wounded, Galois was abandoned on the field of honor. Only hours later a passing peasant took him to a hospital. The only person of his family to be notified before his impending death was his younger brother; on the brink of death, Galois still found strength to tell him, "Don't cry; I need all my courage to die at twenty."

Joseph Liouville was the person who finally "sorted out the mess" and revealed to the world the wonderful results that had been lying dormant in Galois's estate. In 1846, Liouville published all the mathematical papers left by Galois, including the last letter—a mere 64 printed pages. H. Weyl, one of the greatest mathematicians of the twentieth century, described the letter to Chevalier in these words:

> This letter, if judged by the novelty and profundity of ideas it contains, is perhaps the most substantial piece of writing in the whole literature of mankind.

No wonder that Galois is now considered to be one of the founders of modern algebra. It was through his work that the concept of group (which he named) grew to the importance it now has in mathematics. More details about the work of Galois can be found in Edwards 1984, which closely follows Galois's own approach to the theory of equations. For a modern approach to the same material, see Artin 1991. More details about the history of the theory of polynomial equations can be found in van der Waerden 1985. The latest biography of Galois, which contains a detailed account of the duel, is Rigatelli 1996.

4. Arithmetic groups

We must not lose sight of the fact that we are ultimately interested in prime numbers and the factorization of integers. These are arithmetic properties related to the multiplicative structure of the integers. In this section we introduce the groups that will help us in the study of these properties.

Let n be a positive integer. Recall from Chapter 4, section 7, that $U(n)$ is the set of invertible elements of $\mathbb{Z}_n$; that is,

$$U(n) = \{\overline{a} \in \mathbb{Z}_n : \gcd(a, n) = 1\}.$$

Let's show that this set is a group for the operation of multiplication of classes in $\mathbb{Z}_n$.

First of all, we know that the product of two elements of $\mathbb{Z}_n$ is an element of $\mathbb{Z}_n$. Does the same hold for elements of $U(n)$? In other words, is it true that the product of two elements of $U(n)$ is an element of $U(n)$? If not, the multiplication of classes will not be an operation of $U(n)$ in the sense of section 1.

In other words, we must check whether the product of two invertible classes of $\mathbb{Z}_n$ is also an invertible class of $\mathbb{Z}_n$. We have already done this in Chapter 4, section 7, but the argument is so simple, and the result so important, that we may as well repeat it here. Suppose that $\overline{a}$ and $\overline{b}$ are elements of $U(n)$ and that their inverses are $\overline{a'}$ and $\overline{b'}$, respectively. Then $\overline{ab}$ is invertible and its inverse is $\overline{a'b'}$. To check that this is so, it is enough to multiply the two elements

$$\overline{ab} \cdot \overline{a'b'} = \overline{aa'} \cdot \overline{bb'} = \overline{1}.$$

Now we have a set $U(n)$, where an operation is defined, multiplication of classes modulo n. We must show that this operation satisfies the required properties. Associativity is easy, because we already know that multiplication in $\mathbb{Z}_n$ is associative. The identity element is $\overline{1}$. It is an invertible element of $\mathbb{Z}_n$, so it belongs to $U(n)$. That every element of $U(n)$ has an inverse follows from the definition of $U(n)$. So the set $U(n)$ under the operation of multiplication is indeed a group.

That the group $U(n)$ has finite order is clear; it is a subset of $\mathbb{Z}_n$, which is a set of n elements. But for the applications in later chapters we need to know the order of $U(n)$ exactly. Indeed, the order of $U(n)$ is so ubiquitous that it has a special name: $\phi(n)$. Thus we have a function ϕ that, to each positive integer n, associates the number of elements of the set $U(n)$. This is called *Euler's function* or the *totient function*.

We want to find a general formula for $\phi(n)$, but we begin with some special cases. Suppose that p is a positive prime. Then every positive integer smaller than p is prime to p. Hence

$$U(p) = \mathbb{Z}_p \setminus \{\overline{0}\}$$

has $p - 1$ elements. Thus $\phi(p) = p - 1$.

It is also easy to compute $\phi(p^k)$, where p is a positive prime. All we have to do is count the non-negative integers, smaller than p^k, whose greatest common divisor with p^k is 1. But p is prime, so $\gcd(a, p^k) = 1$ if and only if p does not divide a. Hence, it is enough to count the non-negative integers smaller than p^k that *are not* divisible by p. However, it is easier to count those that *are* divisible. Indeed, if $0 \le a < p^k$ is divisible by p, then

$$a = pb \quad \text{where} \quad 0 \le b < p^{k-1}.$$

Thus, there are p^{k-1} non-negative integers, smaller than p^k, that *are* divisible by p. Since there are p^k non-negative integers smaller than p^k, we conclude that there are $p^k - p^{k-1}$ whole numbers in the same interval that are *not* divisible by p. Therefore,

$$\phi(p^k) = p^k - p^{k-1} = p^{k-1}(p - 1).$$

In order to get the general formula we must prove the following result.

Theorem. *If m and n are positive integers such that* $\gcd(m, n) = 1$, *then*

$$\phi(mn) = \phi(m)\phi(n).$$

Before we go into the proof, it should be noted that the hypothesis that m and n are co-prime is necessary. For example, if $m = n = p$, then

$$\phi(mn) = \phi(p^2) = p(p - 1) \qquad \text{but} \qquad \phi(m)\phi(n) = \phi(p)^2 = (p - 1)^2.$$

The proof of the theorem uses the geometric interpretation of the Chinese remainder theorem; that is, the table in Chapter 7, section 3; let's recall how it is constructed. We begin with two positive integers m and n such that $\gcd(m, n) = 1$. Next, we draw a table with m columns and n rows. Each column is indexed by a non-negative integer smaller than m, and each row by a non-negative integer smaller than n. We will think of these numbers as classes in $\mathbb{Z}_m$ and $\mathbb{Z}_n$, respectively. Thus we have a table with mn cells. The entry of the cell in the intersection of column a and row b will be the integer x that satisfies

$$x \equiv a \quad (\mathrm{mod}\ m)$$
$$x \equiv b \quad (\mathrm{mod}\ n),$$

and $0 \le x \le mn - 1$. The integers a and b are the *coordinates* of x. Since m and n are co-prime, it follows from the Chinese remainder theorem that the entry of every cell is uniquely defined by the conditions above. We will think of x as a class in $\mathbb{Z}_{mn}$.

Proof of the theorem. Suppose that $\bar{x} \in \mathbb{Z}_{mn}$, where $0 \le x \le mn - 1$. Let a and b be the coordinates of x in the table constructed above. We begin by proving the following claim.

Claim: $\bar{x} \in U(mn)$ if and only if $\bar{a} \in U(m)$ and $\bar{b} \in U(n)$.

Suppose first that $\bar{x} \in U(mn)$. Then $\bar{x}$ has inverse $\bar{x}' \in U(mn)$, so that $xx' \equiv 1 \pmod{mn}$. But this congruence holds if and only if $xx' - 1$ is divisible by mn. In particular, $xx' - 1$ must be divisible by m; hence $xx' \equiv 1 \pmod m$. But, by definition, $x \equiv a \pmod m$. Thus $ax' \equiv 1 \pmod m$, so that $\bar{a}$ is invertible in $\mathbb{Z}_m$. A similar argument shows that $\bar{b}$ is invertible in $\mathbb{Z}_n$.

In order to prove the converse, suppose that $\bar{x}$ is an element of $\mathbb{Z}_{mn}$ whose coordinates satisfy $\bar{a} \in U(m)$ and $\bar{b} \in U(n)$. We wish to show that $\bar{x}$ is an invertible element of $\mathbb{Z}_{mn}$. By hypothesis, $\bar{a}$ has an inverse $\overline{a'}$ in $\mathbb{Z}_m$, and $\bar{b}$ has an inverse $\overline{b'}$ in $\mathbb{Z}_n$. If $\bar{x} \in \mathbb{Z}_{mn}$ has an inverse, then it must be found somewhere in the table. What are its coordinates? It is reasonable to expect that the coordinates will be the classes $\overline{a'}$ and $\overline{b'}$, which have just been defined. Thus, let $0 \le y \le mn - 1$ be an integer such that

$$y \equiv a' \pmod m$$
$$y \equiv b' \pmod n.$$

Let's prove that $\bar{y} \in \mathbb{Z}_{mn}$ is the inverse of $\bar{x}$. Since $x \equiv a \pmod m$ and $y \equiv a' \pmod m$, we have

$$xy \equiv aa' \equiv 1 \pmod m.$$

Hence $xy - 1$ is divisible by m. A similar argument shows that $xy - 1$ is divisible by n. But $\gcd(m, n) = 1$, so, by the lemma of Chapter 2, section 6, $xy - 1$ is divisible by mn. In other words,

$$\bar{x} \cdot \bar{y} = \bar{1} \quad \text{in} \quad \mathbb{Z}_{mn},$$

which concludes the proof of the claim.

Using the claim, it is easy to prove the theorem. We want to compute $\phi(mn)$. By definition this is equal to the number of elements of $U(mn)$. Hence, by the claim, we must count the number of entries in the table whose first coordinate belongs to $U(m)$ and whose second coordinate belongs to $U(n)$. But $U(m)$ has $\phi(m)$ elements, and $U(n)$ has $\phi(n)$ elements. So the number of those entries is $\phi(m)\phi(n)$. Therefore, $\phi(mn) = \phi(m)\phi(n)$, and the theorem is proved.

We are now ready to find a formula for $\phi(n)$ for any given positive integer n. First of all, we must factor n:

$$n = p_1^{e_1} \dots p_k^{e_k},$$

where $0 < p_1 < \dots < p_k$ are distinct primes. By the theorem

$$\phi(n) = \phi(p_1^{e_1}) \dots \phi(p_k^{e_k}).$$

Using the formula for ϕ of a prime power, deduced above, we obtain

$$\phi(n) = p_1^{e_1-1} \dots p_k^{e_k-1}(p_1 - 1) \dots (p_k - 1).$$

For example, if $n = 120 = 8 \cdot 3 \cdot 5$, then

$$\phi(120) = 2^2(2-1)(3-1)(5-1) = 32.$$

Note that in order to apply the formula we must first be able to factor n completely. This is bad news, if we really need to know $\phi(n)$ for a large integer n. However, as we will see in Chapter 11, it is just this difficulty that makes the RSA cryptosystem secure.

5. Subgroups

If a group H is a subset of a group G and if they share the same operation, we say that H is a *subgroup* of G. Since this definition is very important to what follows, we will carefully dissect it below.

Let G be a group, and denote its operation by $\star$. A non-empty subset H of G is a *subgroup* of G if

(1) $a \star b \in H$, whenever $a, b \in H$;
(2) the identity element of G belongs to H; and
(3) if $a \in H$, then its inverse a' is also an element of H.

In the terminology of section 1, condition (1) says that $\star$ (which is the operation for G) is also an operation for the set H.

Let's begin with an example in the group $\mathbb{Z}$ with addition. Let n be a positive integer, and denote by N the set of all multiples of n, both positive and negative. Is N a subgroup of $\mathbb{Z}$? First, if two integers are multiples of n, so is their sum; so (1) is verified. The identity element of $\mathbb{Z}$ is 0, which is a multiple of n; so $0 \in N$. Finally, $-a \cdot n = (-a) \cdot n \in N$. So the inverse of any element of N also belongs to N. Thus N is indeed a subgroup of $\mathbb{Z}$. This example will reappear in section 8.

Among the examples of section 1 there are several subgroups. Thus, under addition, $\mathbb{Z}$ is a subgroup of $\mathbb{Q}$, which is a subgroup of $\mathbb{R}$, which is a subgroup of $\mathbb{C}$. Under multiplication, the set of non-zero rational numbers is a subgroup of the set of non-zero real numbers, which is a subgroup of the set of non-zero complex numbers. Note that any group has at least two subgroups: the whole group, and the subgroup whose only element is the identity element.

On the other hand, $\mathbb{Q}\backslash\{0\}$ is a group under multiplication, and it is contained in $\mathbb{Q}$, which is a group under addition. However, in this case we do not say that $\mathbb{Q} \backslash \{0\}$ is a subgroup of $\mathbb{Q}$, because they do *not* share the same operation.

Finite groups are very interesting. The fact that a finite set is a group implies that there exist many unexpected relations between their orders. These make it easier to find the subgroups of a group. We will study the simplest of these relations, called *Lagrange's theorem*. By the way, J. L. Lagrange died a year after Galois was born, and his work on the theory of polynomial equations greatly influenced Galois's own work on the subject. Lagrange also contributed to various other branches of mathematics, such as number theory and mechanics.

Lagrange's theorem. *In a finite group, the order of any subgroup divides the order of the whole group.*

Let's make clear what Lagrange's theorem says and especially what it does not say. Suppose that G is a finite group and let H be a subset of G. Then, clearly H has fewer elements than G. Lagrange's theorem says that, if, moreover, H is a subgroup of G, then the order of H actually divides the order of G. This severely restricts those subsets that can be subgroups. However, it is *not* true that if k is a divisor of the order of G, then G must have a subgroup of order G.

The proof of this theorem will be given in section 8. We must first understand the full import of the theorem, and this can only be gauged by looking at some applications. Let's look at D_3, the group of symmetries of an equilateral triangle, which has order 6. Thus, by Lagrange's theorem, it can only have subgroups of orders 1, 2, 3, and 6, which are the factors of 6. Since every subgroup must contain the identity element, the only possible subgroup with one element is $\{e\}$. It is also clear that the only possible subgroup of order 6 of D_3 is D_3 itself. Thus we are left with the task of finding the subgroups of orders 2 and 3 of D_3. This will be done after the next section, where we consider a systematic way of computing subgroups of a group.

6. Cyclic subgroups

Let G be a *finite* group, and denote by $\star$ its operation. Let a be an element of G. We write

$$a^k = a \star a \star \cdots \star a \quad (k \text{ times}).$$

This is the kth *power* of a. Now consider the set of powers of a:

$$H = \{e, a, a^2, a^3, \dots\}.$$

Apparently this is an infinite set. We say *apparently* because $H \subseteq G$, and G is a finite set, so H must also be finite. But this can happen only if there exist powers of a that are equal, even though their exponents are different. In other words, there must be positive integers $n > m$, such that $a^m = a^n$.

Let a' be the inverse of a in G. Multiplying both sides of $a^m = a^n$ with $(a')^m$, we obtain $a^{n-m} = e$, the identity element. Hence, given an element $a \in G$ there exists a positive integer k such that $a^k = e$. Thus

$$a \star a^{k-1} = a^k = e;$$

so the inverse of a is a^{k-1}, which is also a power of a. In particular, the inverse of a belongs to H. Since in multiplying two powers of a we obtain a power of a, we already know enough to conclude that H is a subgroup of G.

What's the order of H? Suppose that k is the *smallest* positive integer for which $a^k = e$. If $n > k$, then we can divide n by k, so that $n = kq + r$ and $0 \le r \le k - 1$. Therefore,

$$a^n = a^{kq+r} = (a^k)^q \star a^r.$$

But $a^k = e$, so $a^n = a^r$. In other words, every power of a whose exponent is greater than k is equal to a power to a smaller exponent. Thus

$$H = \{e, a, a^2, \ldots, a^{k-1}\}.$$

Moreover, all these elements are distinct. For if $r \leq s < k$ and $a^r = a^s$, then $a^{s-r} = e$. Since $s - r < k$, we must have $s - r = 0$; that is, $r = s$. We conclude that the order of H is k.

Thus we have a simple method for constructing subgroups of a given finite group G: Choose any element $a \in G$, then

- The set H of the powers of a in G is a subgroup of G.
- The order of H is equal to the smallest positive integer k for which $a^k = e$.

It is convenient to introduce the following terminology. If the subgroup H is equal to the set of powers of an element a, then H is a *cyclic* subgroup of G, and a is a *generator* of H. The smallest positive integer k for which $a^k = e$ is the *order* of a. From the discussion above we conclude that the order of a is equal to the order of the cyclic subgroup generated by a.

As a simple application, we can determine the structure of a group G whose order is a prime number p. An example of such a group is $\mathbb{Z}_p$ with the operation of addition. Suppose that H is any subgroup of G. By Lagrange's theorem, the order of H must divide the order of G, which is p. Since p is prime, the order of H is either 1 or p. In the first case, $H = \{e\}$, in the second $H = G$. Thus we have found all the subgroups of G. Now choose $a \neq e$ in G and let H be the cyclic subgroup generated by a. Since $e \neq a \in H$, it follows from the argument above that $H = G$. In particular, G is cyclic and any element of G, apart from e, is a generator. These results are summed up in the next theorem.

Theorem (groups of prime order). *If G is a group of prime order, then*

- *G is cyclic;*
- *G has only two subgroups, G itself and $\{e\}$; and*
- *every element of G, except e, generates the whole group.*

Thus every group of prime order is cyclic, but the converse is not true. For example, $U(5)$ has order $\phi(5) = 4$, but it is cyclic, and $\bar{2}$ is one of its generators. We will return to this example in Chapter 10, where we prove the primitive root theorem.

Although we have only discussed in detail cyclic subgroups, it is *not* true that every subgroup of a group is cyclic. This will be clear from the examples of the next section.

7. Finding subgroups

Let's apply the results of the previous section to find all the subgroups of D_3. We saw in section 5 that it is enough to determine the subgroups of orders 2 and 3 of D_3. But 2 and 3 are prime numbers, so these subgroups must be cyclic by the theorem of section 6. Moreover, a cyclic subgroup is completely determined by its generator. Thus it is enough to find which elements of D_3 have order 2, and which have order 3.

Since $\rho^2 \neq e$ and $\rho^3 = e$, it follows that ρ has order 3. We also have

$$(\rho^2)^2 = \rho^3 \rho = \rho \qquad \text{and} \qquad (\rho^2)^3 = (\rho^3) = e,$$

so ρ^2 also has order 3. Since each reflection is its own inverse, the three reflections must have order 2. The cyclic subgroup generated by ρ is

$$R = \{e, \rho, \rho^2\},$$

and it coincides with the cyclic subgroup generated by ρ^2. Hence R is the only subgroup of order 3 of D_3, which is generated by both ρ and ρ^2. Each reflection gives rise to a subgroup of D_3 of order 2, namely

$$\{e, \sigma_1\}, \ \{e, \sigma_2\} \quad \text{and} \quad \{e, \sigma_3\}.$$

Thus we have proved that, apart from $\{e\}$ and D_3 itself, these are the only subgroups of D_3. Does it follow that D_3 has only cyclic subgroups? Not really, because the group D_3 itself is not cyclic. Indeed, if D_3 were cyclic, its generator would have to be an element of order 6. But as we have seen, every element of D_3 has order 1, 2, or 3. However, it is true that every proper subgroup of D_3 is cyclic—a subgroup H of a group G is *proper* if $H \neq G$.

Next, we want an example of a group that is not cyclic, which also contains a proper non-cyclic subgroup. The group we consider is

$$U(16) = \{\overline{1}, \overline{3}, \overline{5}, \overline{7}, \overline{9}, \overline{11}, \overline{13}, \overline{15}\}$$

with multiplication modulo n. This group has order $\phi(16) = 8$. By Lagrange's theorem it can only have subgroups of order 1, 2, 4, or 8. The subgroups of order 1 and 8 are $\{\overline{1}\}$ and $U(16)$, respectively.

To find the cyclic subgroups of orders 2 and 4, we must compute the order of each element of $U(16)$. One quickly finds that $\overline{7}$, $\overline{9}$, and $\overline{15}$ have order 2, and that $\overline{3}$, $\overline{5}$, $\overline{11}$, and $\overline{13}$ have order 4. Thus $U(16)$ has no element of order 8. In particular, it is not a cyclic group.

Are all proper subgroups of $U(16)$ cyclic? Recall that a subgroup of prime order must be cyclic. Thus the order of a non-cyclic proper subgroup of $U(16)$ must be a composite number smaller than 8 that is also a divisor 8. Hence, if such a subgroup exists, its order must be 4. Moreover, if the subgroup is not cyclic, then it cannot have an element of order 4. By Lagrange's theorem, this implies that, apart from $\overline{1}$, all the elements of such a subgroup must have order 2. But $U(16)$ has exactly three elements of order 2 that, together with the identity element, produce a set of four elements, namely

$$\{\overline{1}, \overline{7}, \overline{9}, \overline{15}\}.$$

One easily checks that this is a subgroup of $U(16)$. Thus $U(16)$ has a proper non-cyclic subgroup of order 4.

We can also use the results of the previous section to generalize Fermat's theorem to non-prime moduli.

Euler's theorem. *Let n and a be two integers. If $n > 0$ and $\gcd(a, n) = 1$, then*

$$a^{\phi(n)} \equiv 1 \pmod{n}.$$

The proof is an immediate consequence of Lagrange's theorem. Since a and n are co-prime, it follows that $\overline{a} \in U(n)$. By Lagrange's theorem, the order of $\overline{a}$ divides the order of $U(n)$, which is $\phi(n)$. Denoting by k the order of $\overline{a}$, we have $\phi(n) = kr$ for some positive integer r. Thus

$$(\overline{a})^{\phi(n)} = (\overline{a}^k)^r = \overline{1},$$

which is equivalent to the congruence of Euler's theorem.

8. Lagrange's theorem

Recall the statement of the theorem.

Lagrange's theorem. *In a finite group, the order of any subgroup divides the order of the whole group.*

We begin by defining the equivalence relation used in the proof of the theorem. Let G be a group with operation $\star$ and let H be a subgroup of G. Two elements x and y of G are *congruent modulo H* if

$$x \star y' \in H,$$

where y' is the inverse of y in G. If this is the case, we write $x \equiv y \pmod{H}$.

An example of this relation is the congruence modulo n defined in Chapter 4. Let G be the group $\mathbb{Z}$ with addition, and let H be the set of all multiples of n (both positive and negative multiples). Since the operation in $\mathbb{Z}$ is addition, we have $y' = -y$. Thus, by definition, $x \equiv y \pmod{H}$ if and only if $x - y \in H$, which means that $x - y$ is a multiple of n. Therefore, in this example, $x \equiv y \pmod{H}$ is equivalent to $x \equiv y \pmod{n}$.

Let's now return to the general case of the congruence modulo H in a group G; we must check that it satisfies the three properties that characterize an equivalence relation. Let $x, y, z \in G$.

Reflexive property: We must show that $x \equiv x \pmod{H}$. However, by definition, this holds if $x \star x' \in H$, which, in its turn, follows from the facts that $x \star x' = e$ and that H is a subgroup.

Symmetric property: If $x \equiv y \pmod{H}$, then, by definition, $x \star y' \in H$. But the inverse of an element of a subgroup also belongs to that subgroup. Thus $y \star x'$, the inverse of $x \star y'$, belongs to H. However, $y \star x' \in H$ implies that $y \equiv x \pmod{H}$, which shows that the property holds.

Transitive property: Suppose that $x \equiv y \pmod{H}$ and that $y \equiv z \pmod{H}$. These two congruences are equivalent to $x \star y' \in H$ and $y \star z' \in H$, respectively. Since H is a subgroup,

$$x \star z' = (x \star y') \star (y \star z') \in H.$$

Thus $x \equiv z \pmod{H}$, and the property is proved.

Thus, the congruence modulo H is reflexive, symmetric, and transitive, so it is an equivalence relation. Note that there is an exact correspondence between the conditions that make H a subgroup of G and the properties that make the congruence modulo H an equivalence relation. Lagrange's theorem depends on the subtle balance of all these facts.

Now that we know the congruence modulo H is an equivalence relation, let's find the equivalence class of an element $x \in G$ under this relation. By definition, the equivalence class of x is

$$\{y \in G : y \equiv x \pmod{H}\}.$$

But $y \equiv x \pmod{H}$ is the same as $y \star x' \in H$. Thus $y = h \star x$ for some $h \in H$. Therefore, the equivalence class of x can be written in the form

$$\{h \star x : h \in H\}.$$

This suggests the notation $H \star x$ for this class. Note that the class of the identity element e is H itself. We are now ready to prove Lagrange's theorem.

Proof of Lagrange's theorem. Let G be a finite group, and denote by $\star$ its operation. Let H be a subgroup of G. We must first count the elements of an equivalence class modulo H. If $x \in G$, its equivalence class is

$$H \star x = \{h \star x : h \in H\}.$$

We will show that $H \star x$ has as many elements as H. Since the elements of $H \star x$ are obtained by multiplying the elements of H by a fixed element x of G, it is clear that $H \star x$ cannot have more elements than H. Now suppose that $h_1, h_2 \in H$. If $h_1 \star x = h_2 \star x$, then

$$h_1 = (h_1 \star x) \star x' = (h_2 \star x) \star x' = h_2,$$

where x' is the inverse of x. It follows that distinct elements of H give rise to distinct elements of $H \star x$, when multiplied by x on the right-hand side. Thus the number of elements of $H \star x$ is the same as that of H.

The proof is obtained by putting together these various facts in a coordinated fashion. First, since the congruence modulo H is an equivalence relation, G is the union of the equivalence classes. Suppose that we are considering only distinct classes when we speak of this union. Since distinct classes are necessarily disjoint, it follows that the order of G is equal to the sum of the number of elements of each class. But all the classes have the same number of elements, namely, the order of H. Thus the order of G equals the order of H times the number of (distinct) equivalence classes. In particular, the order of H divides the order of G.

One last comment. As we remarked in section 5, the converse of Lagrange's theorem is false. In other words, if G is a group of order n, and k is a factor of n, then it is *not* necessarily true that G has a subgroup of order k. For example, the symmetry group of a regular tetrahedron has order 12, but it does not have a subgroup of order 6; for a proof see exercises 20, 21, and 22. However, if k is a prime that divides the order of G, then G must have a subgroup of order k.

This is called *Cauchy's theorem*, and it is proved, for instance, in Rotman 1984, theorem 4.2, p. 56.

9. Exercises

1. The group D_4 of symmetries of the square has order 8.
  (1) Write each element of D_4 as a permutation of the vertices of the square.
  (2) Find the inverses of each element of D_4.
  (3) Let ρ be the counterclockwise rotation of $90°$ and σ one of the reflections of the square. Show that $\sigma\rho = \rho^3\sigma$.
  (4) Compute the multiplication table of D_4.

2. Let G be a group. Show that if the square of every element of G is equal to the identity element, then the group is abelian.

3. Compute $\phi(125)$, $\phi(16200)$, and $\phi(10!)$.

4. Let n be a positive integer and let p be a prime factor of n.
  (1) Show that $p - 1$ always divides $\phi(n)$.
  (2) Show that p need *not* divide $\phi(n)$.
  (3) Show that $n > \phi(n)$.

5. Find the values of n for which $\phi(n) = 18$. Do the same for $\phi(n) = 10$ and $\phi(n) = 14$.

6. Show that if $\phi(n)$ is a prime number, then $n = 3$, 4, or 6.

7. Let k be a positive integer. As one can see from exercises 5 and 6, solving the equation $\phi(n) = k$ can be very time-consuming. However, there is a relatively simple algorithm for solving $n\phi(n) = k$, which we now describe. Let $k = n\phi(n)$, and suppose that p is the largest prime that divides k. Show that
  (1) the largest prime factor of n is less than or equal to p;
  (2) the multiplicity of p in the factorization of $n\phi(n)$ must be odd.
It follows from (2) that if the multiplicity of p in the factorization of k is even, then $k = n\phi(n)$ does not have a solution. Suppose that p has odd multiplicity in the factorization of k. Assuming that a solution to the equation exists, and that $n = p^e c$, with $\gcd(c, p) = 1$, we have

$$k = n\phi(n) = p^{2e-1}(p - 1)c\phi(c).$$

This last equation can be used to compute e, since we know the multiplicity of p in the factorization of k. Once e is found, we can write

$$\frac{k}{p^{2e-1}(p - 1)} = c\phi(c).$$

The same method can now be used to find the largest prime that divides c, and so on. Why does this procedure stop?

8. Show that if n is a positive integer and $\phi(n) = n - 1$, then n is prime.

9. Writing n in the form $n = 2^k r$, where r is an odd number, show that if $\phi(n) = n/2$, then n is a power of 2.

10. Show that if m divides n, then $\phi(mn) = m\phi(n)$.

11. Find all the subgroups of the group D_4 of symmetries of the square.

12. Show that $U(2)$ and $U(4)$ are cyclic groups, but that $U(8)$ is not cyclic.

13. Suppose that G is a finite *cyclic* group of order n. Show that if m divides n, then G has an element of order m. Explain why this implies that the converse of Lagrange's theorem holds for *cyclic groups*.

14. Consider the group $U(20)$.

    (1) Compute the order of $U(20)$.
    (2) Compute the order of each element of $U(20)$.
    (3) Show that $U(20)$ is not cyclic.
    (4) Find all the subgroups of order 4 of $U(20)$.
    (5) Find a non-cyclic subgroup of $U(20)$.

15. Let G be a finite group and let S_1 and S_2 be two subgroups of G. Show that:

    (1) $S_1 \cap S_2$ is a subgroup of G.
    (2) If the orders of S_1 and S_2 are co-prime, then $S_1 \cap S_2 = \{e\}$.
    (3) $S_1 \cup S_2$ need not be a subgroup of G.

Hint: To prove (2) recall that $S_1 \cap S_2$ is a subgroup of S_1 and of S_2; use Lagrange's theorem and the fact that the orders are co-prime to show that $S_1 \cap S_2 = \{e\}$. To prove (3) it is enough to give an example in which the union of subgroups is not a subgroup; try $G = D_3$.

16. Let n be an odd, composite, positive integer. Consider the following subset of $U(n)$.

$$H(n) = \{\bar{b} \in U(n) : n \text{ is a pseudoprime to base } b\}$$

Which of the following statements are true, and which are false?

    (1) $H(n)$ is a subgroup of $U(n)$.
    (2) $H(n)$ cannot be equal to $U(n)$ because n is composite.
    (3) $U(n)$ cannot have an element of order $n-1$ because n is composite.

17. Compute the residue of 7^{9876} modulo 60, and that of $3^{87,654}$ modulo 125.

18. Let $p > 0$ be a prime number and let r be a positive integer. Applying Euler's theorem to p^r, show that p^r is a pseudoprime to base b if and only if $b^{p-1} \equiv 1 \pmod{p^r}$.

19. Use the previous exercise to show that 1093^2 is a pseudoprime to base 2.

20. Let G be a finite group and let H be a subgroup of G. Suppose that the quotient of the order of G by the order of H is 2, and let g be an element of G that is *not* in H.

    (1) Show that $g^2 \notin H \star g$.
    (2) Explain why G is the disjoint union of H and $H \star g$.
    (3) Show that $g^2 \in H$.

21. Let $\mathbb{T}$ be the group of symmetries of the regular tetrahedron. This is a group of order 12.

    (1) Find all the elements of $\mathbb{T}$.
    (2) How many elements of order 3 are there in $\mathbb{T}$?

22. Let $\mathbb{T}$ be the group of symmetries of the regular tetrahedron. The purpose of this exercise is to give a proof that this group does *not* have any subgroup of order 6. The proof will be by contradiction, so suppose that H is a subgroup of order 6 of $\mathbb{T}$. Show that H contains all the elements of order 3 of $\mathbb{T}$, and obtain a contradiction using exercise 21.

Hint: If α is an element of order 3 of $\mathbb{T}$, then $(\alpha^2)^2 = \alpha^3\alpha = \alpha$ belongs to H by exercise 20.

23. Write a program, based on exercise 18, to compute the pseudoprimes to base 2 of the form p^2, where $p < 5 \cdot 10^4$ is a prime number; see exercise 11 of Chapter 6.

24. Write a program that, having as input an integer $k > 0$, computes $\phi(k)$. The program will consist essentially of an algorithm to compute the complete factorization of k into primes. Use this algorithm to find all positive integers k smaller than 10^5 for which $\phi(k) = \phi(k+1)$. It is not known whether there are infinitely many k such that $\phi(k) = \phi(k+1)$.

25. An integer $k > 0$ is a *totient* if the equation $\phi(n) = k$ has a solution. Write a program that implements the algorithm of exercise 7. Which positive integers smaller than 10^5 can be shown to be totients using this algorithm?

9

Mersenne and Fermat

In the first two sections of this chapter we study the classic methods for finding factors of Mersenne numbers and Fermat numbers. However, instead of following the original approach of Fermat and Euler, we make full use of the language and results of the theory of groups, developed in Chapter 8. In this way we can deal with these problems in a simple and elegant fashion. The same methods are applied in section 4 to prove a very efficient primality test for Mersenne numbers.

1. Mersenne numbers

One of the best ways to produce very large primes is to use exponential formulae. The oldest exponential formula for primes is the one named after Mersenne. Let n be a positive integer. Recall that the nth *Mersenne number* is

$$M(n) = 2^n - 1.$$

We have seen that if n is a composite, then $M(n)$ is also composite. For if $n = rs$, then

$$2^n - 1 = (2^r)^s - 1 = (2^r - 1)(2^{r(s-1)} + 2^{r(s-2)} + \cdots + 2^r + 1).$$

Hence $M(r)$ is a factor of $M(n) = M(rs)$. Of course, $M(s)$ is also a factor of $M(n)$.

Thus, if we wish to find primes among Mersenne numbers, we need only look among those of the form $M(p)$, where p is prime. However, it is not true that $M(p)$ is prime for every prime p. In this section we describe a method that can be used to find factors of Mersenne numbers when the exponent is prime, but not too large. The key to the method is a general formula for the factors of $M(p)$ discovered by Fermat. In order to prove this formula we need one more general result about groups.

Key lemma. *Let G be a finite group and denote by $\star$ its operation. Let $a \in G$. A positive integer t satisfies $a^t = e$ if and only if t is divisible by the order of a.*

Proof. Let $s > 0$ be the order of a. If s divides t, then $t = sr$ for some positive integer r, and

$$a^t = (a^s)^r = e.$$

To prove the converse, suppose that $a^t = e$. Since the order of a is the *smallest* positive integer s such that $a^s = e$, then $s \leq t$. Dividing t by s, we obtain

$$t = sq + r \quad \text{where} \quad 0 \leq r < s.$$

Thus

$$e = a^t = (a^s)^q \star a^r = a^r$$

because $a^s = e$. Since $r < s$, this can only happen if $r = 0$.

Let's now return to Mersenne numbers. Suppose that $p \neq 2$ is a *prime* number, and that q is a *prime* factor of $M(p) = 2^p - 1$. Then

$$2^p \equiv 1 \pmod{q}.$$

We will consider this congruence as an equation in the group $U(q) = \mathbb{Z}_q \setminus \{\overline{0}\}$, namely

$$\overline{2}^p = \overline{1}.$$

What is the order of $\overline{2}$ as an element of $U(q)$? It follows from the key lemma and the previous equation that the order of $\overline{2}$ divides p. But p is a prime, so that $\overline{2}$ must have order 1 or p. However, $\overline{2}^1 = \overline{1}$ implies that $\overline{1} = \overline{0}$, a contradiction. Thus $\overline{2}$ has order p in $U(q)$. Since, by hypothesis, $p \neq 2$, it follows that $\overline{2}$ has order p. On the other hand, by Fermat's theorem,

$$\overline{2}^{q-1} = \overline{1} \quad \text{in} \quad U(q).$$

Once again, the key lemma implies that the order of $\overline{2}$ divides $q - 1$. Since $\overline{2}$ has order p, it follows that there exists an integer k such that $q - 1 = kp$.

But we can go further. Indeed, $M(p) = 2^p - 1$ is an odd integer, so its prime factors must also be odd. In particular, q is odd. Hence $q - 1$ is even. Since p is odd, we conclude that, in the formula for $q - 1$, the number k must be even. Hence $q - 1 = 2rp$ for some integer r. We have proved the following result.

Fermat's method. *Let $p \neq 2$ be a prime, and let q be a prime factor of $M(p)$. Then $q = 1 + 2rp$ for some positive integer r.*

Let's use this method to find a factor of $M(11) = 2047$. By the formula, any *prime* factor of $M(11)$ is of the form $q = 1 + 22r$. We must now compute q when $r = 1, 2, \ldots$ and find which (if any) of these numbers are factors of $M(11)$. Before we begin, it is helpful to determine how far we have to go with this search. Recall from Chapter 2, section 2, that if $M(p)$ is a composite number, and $q = 1 + 2rp$ is its *smallest* prime factor, then

$$\sqrt{M(p)} \geq q = 1 + 2rp.$$

Since $\sqrt{M(p)} < 2^{p/2}$, it follows that

$$r < \frac{2^{p/2} - 1}{2p}.$$

When $p = 11$ we obtain $r \leq 2$. Thus, the only possible values for r in this case are 1 and 2. Inserting $r = 1$ in the formula $q = 1 + 22r$ gives $q = 23$. A simple division shows that this is indeed a factor of $M(11) = 2047$. The other prime factor is $89 = 1 + 22 \cdot 4$. It is interesting to consider what would have happened if we had tried to factor $M(11)$ by trial division using the algorithm of Chapter 2, section 2. In that case we would have had to try to divide $M(11)$ by each odd prime smaller than 23 before we could stop. There are 8 such primes.

The history of Mersenne numbers is a mine of curious and eccentric stories. One of the best tells of F. N. Cole's talk at a meeting of the American Mathematical Society in 1903. He proved that

$$M(67) = 193{,}707{,}721 \cdot 761{,}838{,}257{,}287$$

by multiplying the two numbers in total silence. The audience applauded enthusiastically! To this day no one knows how to arrive at these factors in a simple way.

As we mentioned in Chapter 3, many of the largest known primes are Mersenne numbers. Of course there are far better methods than Fermat's to check that a given Mersenne number is prime. The most often used of these is the *Lucas–Lehmer test*, which we will explain in section 4.

2. Fermat numbers

We have seen that if $M(n) = 2^n - 1$ is prime, then n must also be prime. This suggests that we should try to determine the values of n for which $2^n + 1$ is prime. Now, if we assume that $p = 2^n + 1$ is prime, then

(2.1) $$\overline{2}^n = -\overline{1} \quad \text{in} \quad U(p).$$

Therefore

$$\overline{2}^{2n} = \overline{1} \quad \text{in} \quad U(p).$$

Thus, by the key lemma, the order of $\overline{2}$ as an element of $U(p)$ divides $2n$. We must now compute this order exactly. Note that it follows from equation (2.1) that the order of $\overline{2}$ can be neither n nor a divisor of n. Since it divides $2n$, the order must be a multiple of 2. Thus there exists a positive integer r such that the order of $\overline{2}$ is $2r$. Clearly r divides n. Now $\overline{2}^{2r} = \overline{1}$ in $U(p)$ implies that

$$\overline{0} = \overline{2}^{2r} - \overline{1} = (\overline{2}^r - \overline{1})(\overline{2}^r + \overline{1})$$

in $\mathbb{Z}_p$. Since we are assuming that p is prime, we conclude that

$$2^r \equiv 1 \pmod{p} \quad \text{or} \quad 2^r \equiv -1 \pmod{p}.$$

Hence p divides $2^r + 1$ or $2^r - 1$. But $p = 2^n + 1$ and $n \geq r$, so we have a contradiction unless $r = n$, and $\overline{2}$ has order $2n$ in $U(p)$.

Since $p - 1 = 2^n$, it follows from Fermat's theorem that

$$\overline{2}^{2^n} = \overline{1} \quad \text{in} \quad U(p).$$

Hence the order of $\overline{2}$ (which is $2n$) divides 2^n. In particular, n must be a power of 2. Summing up, if $2^n + 1$ is prime, then n is a power of 2.

This is the reason why, when looking for primes, we need only consider numbers of the form $2^{2^k} + 1$; that is, Fermat numbers. As we saw in Chapter 3, Fermat believed that these numbers were always prime. It is true that $F(k)$ is prime when $0 \le k \le 4$, but $F(5)$ is composite. This was shown by Euler in 1730. Ironically, his method closely follows Fermat's own method for finding factors of Mersenne numbers, described in section 1. Let's see how Euler's method works.

Suppose that q is a *prime* factor of $F(k)$. Then

(2.2) $\overline{2}^{2^k} = -\overline{1}$ in $U(q)$

and it follows, by the key lemma, that the order of $\overline{2}$ divides 2^{k+1}. But equation (2.2) also implies that this order cannot be a power of 2 smaller than 2^{k+1}. Hence, the order of $\overline{2}$ in $U(q)$ is 2^{k+1}. However, by Fermat's theorem, the order of $\overline{2}$ divides $q - 1$; thus $q - 1 = 2^{k+1}r$.

Euler's method. *If q is a prime factor of $F(k)$, then there exists a positive integer r such that $q = 1 + 2^{k+1}r$.*

We will use Euler's method to find a factor of $F(5) = 2^{32} + 1$. First of all, any prime factor of $F(5)$ must be of the form $q = 1 + 64r$. Thus we must determine if there is some $q < \sqrt{2^{32} + 1} \le 66,000$, of the form above, that divides $F(5)$. The upper bound on q gives $r < 1031$, an uncomfortably large number. The smallest value of r for which q is prime is $r = 3$, which corresponds to $q = 193$. A computation shows that

$$2^{32} \equiv (2^8)^4 \equiv 63^4 \equiv 108 \pmod{193}.$$

Hence 193 is not a factor of $F(5)$. For $r = 4$ we have $q = 257$, which is also prime, but

$$2^{32} \equiv 1 \pmod{257},$$

so that 257 is not a factor either. The next value of r for which q is prime is $r = 7$, and it gives $q = 449$. But

$$2^{32} \equiv (2^{16})^2 \equiv 431^2 \equiv 324 \pmod{449}$$

and 449 is not a factor. The next prime is $q = 577$, which corresponds to $r = 9$. In this case,

$$2^{32} \equiv 287 \pmod{577},$$

so that 577 is not a factor of $F(5)$. Finally, when $r = 10$, we have $q = 641$, which is a factor of $F(5)$.

Luckily, the factor is relatively small, so it could be found using Euler's method and an electronic calculator; Euler, of course, did all these calculations by hand. Unfortunately, one is not often so lucky. The problem is that $F(k)$ is a doubly exponential function. Thus, even for relatively small values of k, it is necessary to search among so many possible candidates that it is almost impossible to find a factor by Euler's method. However, if you want to find a factor of a given Fermat number, there are far better methods than Euler's;

see Lenstra et al. 1993 and Pomerance 1996. More surprisingly, there is a
very efficient test that determines whether a given Fermat number is prime or
composite, as we will see in the next chapter.

A lot is known about Fermat numbers. For example, complete factorizations
are known for all Fermat numbers with $k \leq 9$, and also for $F(11)$. Moreover,
at least one prime factor of $F(k)$ is known when $k \leq 32$, except for $k =
14, 20, 22, 24, 28$, and 31. Actually, the only Fermat numbers we know to be
prime are $F(0), \ldots, F(4)$, and there is some heuristic evidence that these are
the only ones.

Why is there so much interest in Fermat numbers? There are several reasons,
but certainly their long and eventful history is one of them. They are also a good
source of numbers that are large and difficult to factor. This makes them a good
target on which to test the power of new algorithms. Factorization algorithms,
in turn, often require the computer to carry out simple arithmetic and logic
operations a great many times. Thus running them is often a very effective way
of detecting bugs in newly designed computers.

There is a more theoretical reason why Fermat numbers are interesting. In
1801, Gauss showed that if a regular polygon of n sides can be constructed using
a ruler and compass, then n equals a power of 2 times a *prime* Fermat number.
Thus a regular polygon of 17 sides can be so constructed, because $17 = F(2)$,
but not a regular heptagon, because 7 is not a Fermat number. For more details
see Artin 1991, Chapter 13, section 4.

3. Fermat, again

The largest Fermat number for which a factor is known is $F(23,471)$; the
factor is $5 \cdot 2^{23,473} + 1$. It is also the largest Fermat number we know to be
composite. This is a huge number, and you may be wondering what wonderful
algorithm is this that allows one to factor it. The answer is easy. It is the method
of Euler described in the previous section. However, instead of using the method
as Euler did, we turn it upside-down. Euler began with a given Fermat number
and tried to find a factor. We will begin with a number that has a chance of
being the factor of some $F(m)$, and try to find m.

The algorithm begins by choosing two positive integers k and n, the first of
which must be odd. Then the number $q = k \cdot 2^n + 1$ is constructed. It follows
from Euler's method that if this number divides $F(m)$, then $m \leq n - 1$. But q
divides $F(m)$ if and only if

$$2^{2^m} \equiv -1 \pmod{q}.$$

To work with as large a number as the one mentioned above, we need a
very efficient way of computing the congruences. Since only powers of 2 come
into play, this can be done rather easily because

$$(2^{2^i})^2 = 2^{2^{i+1}}.$$

Now we proceed as follows. First, let $r = 2^{2^5}$ and $i = 5$. The variable i is used
to keep track of the exponent. We begin with $i = 5$ because $F(i)$ is prime for

$i < 5$. The core of the algorithm consists of replacing r by the residue of r^2 modulo q and increasing i by 1 at each loop. The algorithm stops when either $r = q - 1$ or $i = n$. In the first case, q divides $F(i)$; in the second, q is not a factor of any $F(i)$. Recall that if $q = k \cdot 2^n + 1$ divides $F(i)$, then $i \leq n - 1$.

This method, with several clever improvements, was used by G. B. Gostin to find factors of $F(15)$, $F(25)$, $F(27)$, and $F(147)$ (see Gostin 1995). His algorithm was written in C and assembler, with some routines being parallelized. The program first generates several million possible values of q. Next, the values that are divisible by small primes are eliminated. The surviving values are tested in the congruences, as explained above.

Reproduced below is part of a table found in Gostin 1995, p. 394. It consists of the values of m, k, and n, for which $k \cdot 2^n + 1$ is a factor of $F(m)$.

m	k	n
15	17,753,925,353	17
64	17,853,639	67
353	18,908,555	355
885	16,578,999	887
1082	82,165	1084
1225	79,707	1231
1451	13,143	1454
3506	501	3508
6390	303	6393
6909	6021	6912

4. The Lucas–Lehmer test

In this section we present a very good primality test for Mersenne numbers that was first conceived by E. Lucas (of Tower of Hanoi fame) in 1878. The test was improved by D. H. Lehmer in 1932, and it is now called the *Lucas–Lehmer test*.

The key ingredient of the Lucas–Lehmer test is the sequence of positive integers $S_0, S_1, S_2, \ldots$, defined recursively by

$$S_0 = 4 \quad \text{and} \quad S_{k+1} = S_k^2 - 2.$$

We will show first that the integers of this sequence can be written as sums of powers of irrational numbers; see exercise 4 of Chapter 5. Let $\omega = 2 + \sqrt{3}$ and $\varpi = 2 - \sqrt{3}$. We will prove by induction on $n \geq 0$ that

(4.1) $$\omega^{2^n} + \varpi^{2^n} = S_n.$$

Clearly $\omega + \varpi = S_0$. Suppose that $\omega^{2^{n-1}} + \varpi^{2^{n-1}} = S_{n-1}$. Squaring both sides, we obtain

$$\omega^{2^n} + 2(\omega\varpi)^{2^{n-1}} + \varpi^{2^n} = S_{n-1}^2.$$

Since $\omega\varpi = 1$, it follows that

$$\omega^{2^n} + \varpi^{2^n} = S_{n-1}^2 - 2,$$

which is equal to S_n by definition.

The Lucas–Lehmer test. *Let p be a positive prime. The Mersenne number $M(p)$ is prime if and only if $S_{p-2} \equiv 0 \pmod{M(p)}$.*

We will prove only that the condition is necessary; the proof of sufficiency is beyond the means of this book. The proof we present originally appeared in Bruce 1993. Although elementary, it uses irrational numbers in a way that is difficult to justify heuristically. Once one accepts the irrational numbers, the rest of the proof closely follows the pattern of the proofs in the previous sections. To understand where the irrational numbers come from, see Bressoud 1989, chapters 10 and 11.

The proof of the Lucas–Lehmer test will be couched in the language of group theory, but the group in question is far more exotic than $U(p)$. The starting point is the subset $\mathbb{Z}[\sqrt{3}]$ of numbers of the form $a + b\sqrt{3}$, where $a, b \in \mathbb{Z}$. These are real numbers, so they can be added and multiplied. The sum and product of two numbers in $\mathbb{Z}[\sqrt{3}]$ also belong to $\mathbb{Z}[\sqrt{3}]$. Moreover, like $\mathbb{Z}$, the set $\mathbb{Z}[\sqrt{3}]$ is a group under addition, but not under multiplication. These facts are very easily checked. Note that every $a \in \mathbb{Z}$ can be written in the form $a = a + 0\sqrt{3}$, so that $\mathbb{Z} \subseteq \mathbb{Z}[\sqrt{3}]$.

Now let $q \geq 0$ be a prime integer, and write

$$I(q) = \{q\alpha : \alpha \in \mathbb{Z}[\sqrt{3}]\}.$$

Clearly $0 = 0q \in I(q)$. Since $q\alpha + q\beta = q(\alpha + \beta)$, it follows that the sum of two numbers in $I(q)$ is also in $I(q)$. Moreover, for any $\alpha \in \mathbb{Z}[\sqrt{3}]$, both $q\alpha$ and $-q\alpha$ belong to $I(q)$. Thus $I(q)$ is a subgroup of the additive group $\mathbb{Z}[\sqrt{3}]$.

Now, the relation *congruence modulo $I(q)$* is an equivalence relation in $\mathbb{Z}[\sqrt{3}]$, as shown in Chapter 8, section 8. Recall that if $\alpha, \beta \in \mathbb{Z}[\sqrt{3}]$, then $\alpha \equiv \beta \pmod{I(q)}$ when $\alpha - \beta \in I(q)$. See also exercises 10 and 11 of Chapter 4.

Now if $\alpha \in \mathbb{Z}[\sqrt{3}]$, then $\alpha = a_1 + a_2\sqrt{3}$, where $a_1, a_2 \in \mathbb{Z}$. Dividing a_1 and a_2 by q, we have $a_1 = qb_1 + r_1$ and $a_2 = qb_2 + r_2$, with $0 \leq r_1, r_2 < q$. Writing $\rho = r_1 + r_2\sqrt{3}$, it follows that

$$\alpha - \rho = q(b_1 + b_2\sqrt{3}).$$

So $\alpha \equiv \rho \pmod{I(q)}$. The number ρ is called the *reduced form* of α modulo $I(q)$. Since the remainder of the division of integers is unique, each element of $\mathbb{Z}[\sqrt{3}]$ has only one reduced form modulo $I(q)$. Note that there are exactly q^2 distinct reduced forms modulo $I(q)$ in $\mathbb{Z}[\sqrt{3}]$.

Now each equivalence class of $\mathbb{Z}[\sqrt{3}]$ modulo $I(q)$ can be represented by an element in reduced form. Moreover, two classes represented by elements whose reduced forms are different must be distinct. Thus the set $\mathbb{Z}_q[\sqrt{3}]$, of equivalence classes modulo $I(q)$, must have q^2 elements.

The equivalence class of $\alpha \in \mathbb{Z}[\sqrt{3}]$ modulo $I(q)$ will be denoted by $\tilde{\alpha}$. We can define a multiplication in $\mathbb{Z}_q[\sqrt{3}]$ by

$$\tilde{\alpha}\tilde{\beta} = \widetilde{\alpha\beta}.$$

The proof that this definition is independent of the choice of the representatives of the classes is like that of the corresponding result for modular arithmetic; see Chapter 4, section 3. If $\alpha = a_1 + a_2\sqrt{3}$ and $\beta = b_1 + b_2\sqrt{3}$, then a simple calculation shows that $\tilde{\alpha}\tilde{\beta}$ is represented by

$$(a_1 b_1 + 3a_2 b_2) + (a_1 b_2 + a_2 b_1)\sqrt{3}.$$

One readily checks that this multiplication is associative, commutative, and has $\tilde{1}$ as its identity element. However, $\mathbb{Z}_q[\sqrt{3}]$ is not a group under this operation (see exercise 9). As in the case of modular arithmetic, we get around this difficulty by considering the set $V(q)$ of invertible elements of $\mathbb{Z}_q[\sqrt{3}]$. This is a group, because the product of invertible elements in $\mathbb{Z}_q[\sqrt{3}]$ is again invertible, as one easily checks; see Chapter 8, section 4. Note that since

$$V(q) \subset \mathbb{Z}_q[\sqrt{3}] \setminus \{\tilde{0}\},$$

it follows that the order of $V(q)$ is necessarily smaller than q^2. Note also that, since $\omega\varpi = 1$, then both ω and ϖ belong to $V(q)$. We are now ready to prove the Lucas–Lehmer test.

Proof of the Lucas–Lehmer test. Suppose that, for some prime p, the Mersenne number $M(p)$ divides S_{p-2}. By (4.1), there exists an integer r such that

$$\omega^{2^{p-2}} + \varpi^{2^{p-2}} = rM(p).$$

Multiplying this equation by $\omega^{2^{p-2}}$, and recalling that $\omega\varpi = 1$, we obtain

(4.2) $$\omega^{2^{p-1}} + 1 = rM(p)\omega^{2^{p-2}}.$$

Suppose now that $M(p)$ is composite and that q is its smallest prime factor, and let's aim at a contradiction. Since q divides $M(p)$, it follows from (4.2) that

(4.3) $$\tilde{\omega}^{2^{p-1}} = -\tilde{1}$$

in $\mathbb{Z}_q[\sqrt{3}]$. Squaring (4.3), we get

$$\tilde{\omega}^{2^p} = \tilde{1}.$$

Now it follows from the key lemma that the order of $\tilde{\omega}$ divides 2^p, but equation (4.3) also tells us that this order cannot be a power of 2 smaller than 2^p. Hence, the order of $\tilde{\omega}$ in $V(q)$ is 2^p. But, by Lagrange's theorem, the order of $\tilde{\omega}$ divides the order of $V(q)$. Since $V(q)$ has an order smaller than or equal to $q^2 - 1$, we have $2^p \leq q^2 - 1$. But q is the smallest prime divisor of $M(p)$, so $q^2 \leq M(p)$. Thus

$$2^p \leq q^2 - 1 < 2^p - 1,$$

which is a contradiction. Thus, if $M(p)$ divides S_{p-2}, then $M(p)$ is prime. This shows that the condition is necessary. For a proof of the sufficiency see Bressoud 1989, theorem 11.10, p. 175.

Although the test is hard to prove, it is very easy to use and implement. In 1978 two high school students, Laura Nickel and Curt Noll, used the Lucas–Lehmer test in their local university mainframe to show that $M(21,701)$ is prime. Their feat made its way to the front page of the *New York Times*. This test is also the basis of GIMPS, the *Great Internet Mersenne Prime Search*[1], which offers free software (and source code) through the Web, so that any owner of a personal computer can enter the search for big prime numbers. The largest known Mersenne prime, the $2,098,960$-digit number $M(6972593)$ was discovered by Nayan Hajratwala in June 1999 using the GIMPS software.

5. Exercises

1. Let p, q be prime numbers. Show that if the congruence $x^p \equiv 1 \pmod{q}$ has a solution $x \not\equiv 1 \pmod{q}$, then $q \equiv 1 \pmod{p}$.

2. Find all solutions of the congruence $x^{17} \equiv 1 \pmod{43}$.

3. Find the generators of the cyclic group $U(17)$. Use them to solve the congruence $7^x \equiv 6 \pmod{17}$.

4. Use Fermat's method to find prime factors of the Mersenne numbers $M(23)$ and $M(29)$, and to show that $M(7)$ is prime.

5. Use Euler's method to show that $F(4)$ is a prime number.

6. Let $k \geq 2$ be an integer, and $\alpha = 2^{2^{k-2}}(2^{2^{k-1}} - 1)$. Let p be a prime factor of $F(k) = 2^{2^k} + 1$. Show that
   (1) $\alpha^2 \equiv 2 \pmod{p}$,
   (2) α has order 2^{k+2} modulo p, and
   (3) $p = 2^{k+2}r + 1$, where r is a positive integer.
Note that this result makes the search for factors by Euler's method a little more efficient.

7. Find the Fermat number $F(k)$ of which $7 \cdot 2^{14} + 1$ is a prime factor.

8. In 1640, Frenicle asked Fermat, through Mersenne, if there were any perfect numbers between 10^{20} and 10^{22}. We know that Frenicle meant even (or Euclidean) perfect numbers; see Chapter 2, exercises 8, 9, and 10. The purpose of this exercise is to give Fermat's proof that there are no even perfect numbers in this range. Recall that we showed in the exercises of Chapter 2 mentioned above that every even perfect number is of the form $2^{n-1}(2^n - 1)$, with $2^n - 1$ a Mersenne prime.

---

[1] You will find the GIMPS software at http://www.mersenne.org.

(1) Show that if $n \geq 2$, then $-1 < \log(1 - 2^{-n}) < 0$, where log stands for logarithms to base 10, and conclude that

$$n \log 2 - 1 < \log(2^n - 1) < n \log 2.$$

(2) Applying logarithms to the inequality

$$10^{20} < 2^{n-1}(2^n - 1) < 10^{22}$$

and using (1), show that $35 \leq n \leq 37$. Don't forget that n is an integer.

(3) Show that $M(n)$ is never a prime when $35 \leq n \leq 37$.

For more details see Weil 1987, Chapter II, section IV.

9. Let $q \geq 0$ be a prime number and denote by $V(q)$ the set of invertible elements of $\mathbb{Z}_q[\sqrt{3}]$.

(1) Show that $\widetilde{0} \notin V(q)$.

(2) Show that, if a is an integer that is not divisible by q, then

$$\widetilde{a}, \ \widetilde{a\sqrt{3}} \in V(q).$$

(3) Find all the elements of $V(5)$.

10. Program Fermat's method to find factors of Mersenne numbers. The input of the program will be a prime $p > 0$, and its output will be the smallest prime factor of $M(p) = 2^p - 1$, or a message stating that $M(p)$ is prime. The program consists basically of an application of the algorithm presented in section 2 of the Appendix for the computation of the residue of 2^n modulo q, where $q = r \cdot 2p + 1$, for some $0 \leq r \leq [(2^{p/2} - 1)/2p]$. If the residue is 1, then q is a factor of $M(p)$. If no factor q is found in the given range, then $M(p)$ is prime. Note that it does not pay to find out whether a given q is or is not prime before calculating the residue. Use this program to find the primes p, between 2 and 300, for which $M(p)$ is prime (see Chapter 3, section 2).

11. Program the algorithm of section 3. It must have as input a prime p, and must search for the value of m for which p divides $F(m)$. Of course, you must take into account that such an m may not exist. Note that if $p = k \cdot 2^n + 1$, then $m < n$. This gives an upper bound beyond which we can be sure that the search will fail. Use this program to find the Fermat numbers that are divisible by $37 \cdot 2^{16} + 1$ and $11,131 \cdot 2^{12} + 1$, respectively. Note that it would not be possible to find these numbers with a home computer if we had to calculate all the digits of the corresponding Fermat numbers.

10

Primality tests and primitive roots

In this chapter we prove Gauss's famous theorem that $U(p)$ is a cyclic group when p is prime. This is the inspiration for a test that, unlike those of Chapter 6, can be used to *prove* that a given integer is prime. It is with this test that we begin the chapter.

1. Lucas's test

Suppose that we wish to determine whether a given positive odd integer n is prime. A possible strategy is to try to show that $U(n)$ has order $n - 1$; in other words, that $\phi(n) = n - 1$. This implies that every positive $a < n$ is co-prime to n, so n must be prime. At first sight, this seems a hopeless task. How can one count the elements of $U(n)$ when n is large? The light at the end of the tunnel is provided by the following theorem, which will be proved in section 5.

Primitive root theorem. *If p is prime, then $U(p)$ is a cyclic group.*

It follows from the theorem that if p is prime, then there exists an element $\bar{b} \in U(p)$, whose order is $p - 1$. In other words,

$$\bar{b}^{p-1} = \bar{1} \quad \text{but} \quad \bar{b}^r \neq \bar{1}$$

if $r < p - 1$. This suggests the following strategy. Suppose that an odd integer $n > 0$ is given, and that we somehow find $\bar{b} \in U(n)$ with order $n - 1$. By Lagrange's theorem the order of $\bar{b}$ must divide the order of $U(n)$. Hence $n - 1$ must divide $\phi(n)$. However, $\phi(n) \leq n - 1$, so that $\phi(n) = n - 1$. Thus n is prime. According to the *primitive root theorem*, such a b always exists if n is prime. But this is not to say that it is going to be easy to find it; for that we will need a good deal of luck.

In order to apply this strategy to test the primality of an integer n, we need a simple way of checking that a given element of $U(n)$ has order $n - 1$. The statement of the test given below was proposed by D. H. Lehmer in 1927, and was based on a slightly weaker test first suggested by E. Lucas.

Lucas's test. *Let n be an odd positive integer, and let b be an integer such that $2 \leq b \leq n - 1$. If, for each prime factor p of $n - 1$, we have*

(1) $b^{n-1} \equiv 1 \pmod{n}$ *and*

(2) $b^{(n-1)/p} \not\equiv 1 \pmod{n}$,

then n is prime.

Proof. Let k be the order of $\overline{b}$ in $U(n)$. We want to show that $k = n - 1$. Since $\overline{b}^{n-1} = \overline{1}$, it follows from the key lemma that k divides $n - 1$. Thus there exists an integer $t \geq 1$ such that $n - 1 = kt$. We must show that $t = 1$.

Suppose, by contradiction, that $t > 1$. Then t is divisible by some prime q. But if q divides t, then q divides $n - 1$. Thus, $(n - 1)/q$ and t/q are whole numbers. Moreover, from

$$\frac{n - 1}{q} = k \cdot \frac{t}{q}$$

we conclude that k divides $(n - 1)/q$. Using the key lemma again, we deduce that $\overline{b}^{(n-1)/q} = \overline{1}$; and this contradicts that hypotheses. Therefore, $t = 1$, and so $k = n - 1$. Now, by Lagrange's theorem, the order of $\overline{b}$ must divide the order of $U(n)$. Hence $n - 1$ divides $\phi(n) \leq n - 1$, so $\phi(n) = n - 1$ and n is prime.

The previous tests detected with certainty only that a number was composite. Lucas's test detects with certainty only that a number is prime. Note that to be successful in applying this test we must be able to factor $n - 1$ completely. Luckily, this is often easy to do for certain families of numbers; for example, factors of Fermat numbers. One must also be lucky in the choice of the base b, or the test will have an inconclusive output even though the number is prime.

Let's use Lucas's test to prove a primality test for Fermat numbers first proposed by Jean François Théophile Pepin (1826–1904).

Pepin's test. *The Fermat number $F(k)$ is prime for some $k > 1$ if and only if*

$$5^{(F(k)-1)/2} \equiv -1 \pmod{F(k)}.$$

Suppose first that the congruence above is satisfied, and note that $F(k) - 1$ has 2 as its only prime factor. Since $5^{(F(k)-1)/2} \equiv -1 \not\equiv 1 \pmod{F(k)}$, while

$$5^{F(k)-1} \equiv (5^{(F(k)-1)/2})^2 \equiv (-1)^2 \equiv 1 \pmod{F(k)},$$

it follows that $F(k)$ is prime by Lucas's test. The converse is harder to prove; it depends on the *law of quadratic reciprocity*, and we will not prove it here. See Hardy and Wright 1994, Chapter VI, for a proof.

Let's show that $F(4)$ is prime using Pepin's test. We have

$$\frac{F(4) - 1}{2} = 2^{15},$$

but

$$5^{2^{15}} \equiv 5^{32,768} \equiv 65,536 \equiv -1 \pmod{F(4)}.$$

Thus the condition of the test is verified, and $F(4)$ is prime. Pepin's test is easy to apply because one has only to compute squares modulo $F(k)$, and that can be done quite rapidly.

As a second example, let's prove that the rep-unit

$$R(19) = \underbrace{1,111,111,111,111,111,111}_{19}$$

is prime. In order to apply Lucas's test we must first find the complete factorization of $R(19) - 1$, which is

$$R(19) - 1 = 2 \cdot 3^2 \cdot 5 \cdot 7 \cdot 11 \cdot 13 \cdot 19 \cdot 37 \cdot 52{,}579 \cdot 333{,}667.$$

Our first choice will be $b = 2$. Using a computer algebra system, it is easy to show that

$$2^{R(19)-1} \equiv 1 \pmod{R(19)}.$$

But, unfortunately,

$$2^{(R(19)-1)/2} \equiv 1 \pmod{R(19)}.$$

Thus condition (2) of Lucas's test fails for $b = 2$ and $p = 2$. Hence 2 is not a good choice of base.

Next we choose $b = 3$. Using a computer algebra system, we find that

$$3^{R(19)-1} \equiv 1 \pmod{R(19)},$$

so that condition (1) of Lucas's test holds. We must now find the residues of

$$3^{(R(19)-1)/p}$$

modulo $R(19)$ for each one of the primes p that appear in the factorization of $R(19)-1$. These residues, computed with the help of a computer algebra system, are displayed in the table below.

Prime factor p	Residue of $3^{(R(19)-1)/p}$ modulo $R(19)$
2	$R(19) - 1$
3	933,000,903,779,960,656
5	97,919,522,321,038,174
7	742,392,324,159,673,027
11	920,873,402,557,886,628
13	114,592,042,672,083,983
19	10^{11}
37	397,724,716,798,816,350
52,579	760,105,763,664,485,871
333,667	555,602,369,615,218,524

Thus condition (2) of Lucas's test holds for each one of the prime factors of $R(19) - 1$. Hence $R(19)$ is indeed a prime number.

Only five prime rep-units are known. The first, of course, is $R(2) = 11$, and $R(19)$ is the second one; the other ones are $R(23)$, $R(317)$, and $R(1031)$. It is not difficult to prove that $R(23)$ is prime using the improved version of Lucas's test that we describe in the next section (see exercise 4).

2. Another primality test

A simple example will bring to light one of the most obvious difficulties one faces in applying Lucas's test. Suppose we want to show that $n = 41$ is prime using the test. First, we must factor $n - 1 = 40$, which gives $n - 1 = 2^3 \cdot 5$. Thus, it is necessary to find an integer b such that $2 \leq b \leq 40$, which also satisfies the congruences

$$b^{40} \equiv 1 \pmod{41}$$
$$b^{20} \not\equiv 1 \pmod{41}$$
$$b^8 \not\equiv 1 \pmod{41}.$$

We begin trying $b = 2$, but we soon find that $2^{20} \equiv 1 \pmod{41}$. Next we try $b = 3$, but although $3^{20} \equiv 40 \pmod{41}$, it turns out that $3^8 \equiv 1 \pmod{41}$. To make things even more frustrating, $2^8 \equiv 10 \pmod{41}$. Thus different bases satisfy the congruences that come from different prime factors of $n - 1$. The trouble is that Lucas's test requires that one base be used in all the congruences. For $n = 41$, the smallest such base is 7.

In 1975, Brilhart, Lehmer, and Selfridge realized that one could rework Lucas's test so that different bases could be chosen for different prime factors. This makes the test a lot easier to use.

Primality test. *Let $n > 0$ be an odd integer so that*

$$n - 1 = p_1^{e_1} \dots p_r^{e_r},$$

where $p_1 < \dots < p_r$ are positive prime numbers. If, for each $i = 1, \dots, r$ there exist integers b_i $(2 \leq b_i \leq n - 1)$ such that

$$b_i^{n-1} \equiv 1 \pmod{n} \quad and$$
$$b_i^{(n-1)/p_i} \not\equiv 1 \pmod{n},$$

then n is prime.

Note that the b_is need *not* be all distinct.

Proof. Let $i = 1$; the same argument applies to $i = 2, \dots, r$. First, we must compute the order of b_1 in $U(n)$; let's denote it by s_1. It follows from the key lemma and from equation $b_1^{n-1} \equiv 1 \pmod{n}$ that s_1 divides $n - 1$. Hence, the primes that appear in the factorization of s_1 are among the primes $p_1, \dots, p_r$. Thus

$$s_1 = p_1^{k_1} \dots p_r^{k_r},$$

where $k_1 \leq e_1, \dots, k_r \leq e_r$.

On the other hand, we know that $b_1^{(n-1)/p_1} \not\equiv 1 \pmod{n}$. Therefore, $(n - 1)/p_1$ is *not* divisible by s_1. But

$$(n - 1)/p_1 = p_1^{e_1 - 1} p_2^{e_2} \dots p_r^{e_r}.$$

Comparing the factorizations of s_1 and $(n - 1)/p_1$, and keeping in mind that s_1 does not divide $(n - 1)/p_1$, we see that $k_1 = e_1$. In other words, $p_1^{e_1}$ divides s_1.

Recall that s_1 is the order of $\overline{b_1}$ in $U(n)$. By Lagrange's theorem, s_1 divides the order of $U(n)$. Thus s_1 divides $\phi(n)$. Since $p_1^{e_1}$ divides s_1, it follows that $p_1^{e_1}$ divides $\phi(n)$.

A similar argument can be used for $i = 2, \ldots, r$, so the congruences of the test imply that $p_1^{e_1}, p_2^{e_2}, \ldots, p_r^{e_r}$ divide $\phi(n)$. These are pairwise co-prime, because they are powers of distinct primes. Thus, by the lemma of Chapter 6, section 2, the product

$$p_1^{e_1} \ldots p_r^{e_r} = n - 1$$

also divides $\phi(n)$. Since $\phi(n) \leq n-1$, we must have that $\phi(n) = n-1$. Hence n is prime.

Finally, putting together results from various chapters, we end up with an efficient strategy for testing primality. Suppose that a large odd integer $n > 0$ is given. To check whether n is prime we can proceed as follows:

(1) Check whether n is divisible by primes smaller than 5000.
(2) Assuming that n is not divisible by any of these primes, apply Miller's test to n using as bases the first 20 primes.
(3) Assuming that the output of Miller's test to all these bases was "inconclusive", apply the test above to n.

3. Carmichael numbers

We have seen that it is possible to characterize Carmichael numbers in terms of their factorizations into primes. This is Korselt's theorem, a result for which we gave an incomplete proof in Chapter 6, section 2. Since the missing ingredient was the *primitive root theorem*, we can now complete the proof of Korselt's theorem. First, let's recall the statement of the theorem.

Korselt's theorem. *An odd integer $n > 0$ is a Carmichael number if and only if the following conditions hold for each prime factor p of n:*

(1) *p^2 does not divide n, and*
(2) *$p - 1$ divides $n - 1$.*

In Chapter 6, section 2, we saw that if (1) and (2) hold, then n is a Carmichael number. We also showed that if n is a Carmichael number, then (1) must hold. To complete the proof we need only show that, if n is a Carmichael number, then (2) also holds. This is where we use the *primitive root theorem*.

Suppose that n is a Carmichael number. By definition, $b^n \equiv b \pmod{n}$ for every integer b. Let p be a prime factor of n. By the *primitive root theorem* the group $U(p)$ is cyclic, generated by some class $\overline{a}$.

Since n is a Carmichael number, $a^n - a$ is divisible by n. Since p divides n, it follows that p also divides $a^n - a$. Thus $a^n \equiv a \pmod{p}$. But p is a prime that does not divide a, so a is invertible modulo p. Therefore, $a^{n-1} \equiv 1 \pmod{p}$, and it follows from the key lemma that the order of $\overline{a}$ divides $n - 1$. However, the order of $\overline{a}$ is $p - 1$, because it is a generator of $U(p)$. Hence, $p - 1$ divides $n - 1$, and the proof is complete.

4. Preliminaries

In this section we prove a result about the orders of elements in an abelian group that is an important ingredient in the proof of the primitive root theorem.

Lemma. *If an abelian group G has elements of orders r and s, then G has an element whose order is the least common multiple of r and s.*

The proof of the primitive root theorem in the next section will be algorithmic. Thus the proof of the lemma will be as important as its statement, because it is not enough to know that the element exists; we must have a method for finding it.

Proof. As usual, we denote the operation of G by $\star$. Suppose that a and b are elements of G of orders r and s, respectively. We must first factor r and s in terms of prime powers, say

$$r = p_1^{e_1} \ldots p_k^{e_k} \quad \text{and} \quad s = p_1^{f_1} \ldots p_k^{f_k},$$

where $p_1, \ldots, p_k$ are distinct primes. Note that we have written the same primes in the two factorizations. This does not mean that we are assuming that r and s have the same prime factors. If, for instance, one of the primes is not a factor of r, then its multiplicity in the factorization will be zero.

Note also that we will not abide by the previous convention that the primes be listed in increasing order. For the purposes of this proof it is better to assume that the primes have been arranged so that, for some $1 \leq g \leq k$, we have

$$e_1 \geq f_1, \ldots, e_g \geq f_g, \quad \text{but} \quad e_{g+1} < f_{g+1}, \ldots, e_k < f_k.$$

Now write

$$r' = p_1^{e_1} \ldots p_g^{e_g} \quad \text{and} \quad s' = p_{g+1}^{f_{g+1}} \ldots p_k^{f_k}.$$

Note that the prime factors of r' and s' are all distinct, so that $\gcd(r', s') = 1$. On the other hand, $r's'$ is the product of powers of the primes $p_1, \ldots, p_k$. Moreover, the multiplicity of p_i is the largest of the numbers e_i and f_i. Thus, $r's'$ is the least common multiple of r and s.

Since r' divides r and s' divides s, there exist positive integers u and v such that $r = r'u$ and $s = s'v$. We wish to show that if $a \in G$ has order r, and if $b \in G$ has order s, then $c = a^u \star b^v$ is an element of G whose order is the least common multiple of r and s.

Before we prove this, recall that one of the hypotheses of the lemma is that the group must be abelian. This is necessary because we need to know that, if $x, y \in G$, then

$$(x \star y)^q = x^q \star y^q,$$

and this is not true if $\star$ is not commutative. Indeed, the result of the lemma does not hold if the group is not abelian. This is illustrated by an example at the end of this section.

Let m be the least common multiple of r and s, and let n be the order of c. Thus both r and s divide m. Suppose that $m = rt = sq$ for positive integers t and q. Then

$$c^m = (a^u \star b^v)^m = a^{um} \star b^{vm} = (a^r)^{ut} \star (b^s)^{vq} = e,$$

and it follows from the key lemma that n must divide m.

On the other hand, since n is the order of c,

$$e = c^n = a^{un} \star b^{vn}.$$

Hence

$$e = (a^{un} \star b^{vn})^{r'} = (a^{r'u})^n \star b^{vnr'} = b^{vnr'}$$

because $r'u = r$ is the order of a. Thus, by the key lemma, the order of b (which is s) must divide vnr'. Since $s = s'v$, it follows that s' divides nr'. However, r' and s' are co-prime. Therefore, by the lemma of Chapter 2, section 6, s' divides n.

A similar argument shows that r' divides n. But r' and s' are co-prime. Thus by the lemma of Chapter 2, section 6, $r's'$ divides n. However, $r's'$ is the least common multiple of r and s, so m divides n. The first part of the argument showed that n divides m, so $m = n$, and c is an element whose order is the least common multiple of r and s.

One deduces from the proof of the lemma that if r and s are co-prime, then we can choose $r' = r$ and $s' = s$. Thus $u = v = 1$, and $c = a \star b$ has order rs. In general, it is enough to know a and b, and the factorizations of r and s, to compute c.

Finally, we should give an example to show that the hypothesis that G is abelian cannot be removed from the statement of the lemma. We give an example in the group D_3. Recall that this is the group of symmetries of an equilateral triangle, and that it is not abelian. Consider the reflection σ_1 and the rotation ρ. The order of σ_1 is 2, and the order of ρ is 3. If the lemma were true for non-abelian groups, then D_3 would have an element of order 6. But such an element does not exist. Note that if we ignore the fact that D_3 is not abelian and write the element constructed in the proof of the lemma for this example, we get $\sigma_1 \rho = \sigma_3$, which has order 2.

5. Primitive roots

Let $p > 3$ be a prime number. The order of $U(p)$ is $\phi(p) = p - 1$, an even composite number. However, $U(p)$ is a cyclic group. A generator of $U(p)$ is also called a *primitive root*. The use of the word *root* in this context may seem somewhat peculiar, but it is easy to justify. By Fermat's theorem the elements of $U(p)$ are the roots of the polynomial equation $x^{p-1} - \bar{1} = \bar{0}$ with coefficients in $\mathbb{Z}_p$. A generator of $U(p)$ is a primitive root because all the other roots can be obtained as powers of a primitive root. The same phenomenon is present when we solve equations with complex coefficients. For example, the equation $x^p - 1 = 0$ has primitive root $\cos(2\pi/p) + i\sin(2\pi/p)$.

The proof of the existence of a primitive root that we give is constructive. In other words, it actually provides an algorithm to compute a primitive root modulo p. The existence of primitive roots for prime moduli was divined by L. Euler, but the first correct proof of their existence was given by C. F. Gauss in his *Disquisitiones arithmeticæ*; see Gauss 1986, sections 73 and 74.

Primitive root theorem. *If p is a prime, then $U(p)$ is a cyclic group.*

Proof. We can assume that $p \geq 5$, because the theorem is obviously true for $p = 2$ or 3. Choose an element $\overline{a_1} \in U(p)$, where $1 < a_1 < p - 1$. Since the choice is arbitrary, we can always begin with $a_1 = 2$. Let k_1 be the order of $\overline{a_1}$. If $k_1 = p - 1$, we have already found a generator of $U(p)$.

Thus we can assume that $k_1 < p - 1$. The class $\overline{a_1}$ is a solution of $x^{k_1} - 1$ in $\mathbb{Z}_p$. Since p is prime, it follows from the theorem of Chapter 5, section 4, that this polynomial equation cannot have more than k_1 distinct solutions. On the other hand, all the elements of

$$H = \{\overline{1}, \overline{a_1}, \overline{a_1}^2, \ldots \overline{a_1}^{k_1-1}\}$$

are solutions of $x^{k_1} - 1$. Since H has k_1 distinct elements, it must contain all the roots of $x^{k_1} - 1$. But $k_1 < p - 1$, thus there exists an element $\overline{b} \in U(p)$ that does not belong to H. In particular, $\overline{b}$ is not a solution of $x^{k_1} - 1$. Therefore, by the key lemma, the order of $\overline{b}$ does not divide k_1.

Denote by r the order of $\overline{b}$. There are two possible cases. If $r = p - 1$, then $\overline{b}$ generates $U(p)$, and we have proved the theorem. Hence we can assume that $r < p - 1$. By the lemma of section 4, there exists an element $\overline{a_2}$ whose order k_2 is the least common multiple of k_1 and r.

Since r does not divide k_1, it follows that $k_2 > k_1$. Now we need only carry on like this until we obtain an element of order $p - 1$. Note that this process must come to a stop, thus producing the required generator. If it didn't, it would be possible to construct an infinite sequence of positive integers $k_1 < k_2 < k_3 < \ldots$, each of which is the order of an element of $U(p)$. However, these integers would have to be smaller than $p - 1$, and this leads to a contradiction.

This method produces, in a systematic way, a generator of $U(p)$, but not necessarily its smallest generator. Note that the converse of the theorem is false. For example, $U(4)$ is cyclic, even though 4 is composite. Thus the fact that $U(n)$ is cyclic does not imply that n is prime. Indeed, it can be shown that $U(n)$ is cyclic if and only if n is equal to 1, 2, 4, p^k, or $2p^k$, where p is an odd prime number. For a proof of this result see Giblin 1993, Chapter 8.

6. Computing orders

In this section we apply Gauss's method, described in section 5, to find a generator of $U(41)$. But first we must find a simple way to compute the order of an element of $U(p)$.

Let p be an odd prime, and $\overline{a} \in U(p)$. We want to compute the order k of $\overline{a}$. By Lagrange's theorem, k must divide $\phi(p) = p - 1$, which is the order of

$U(p)$. Let's assume that we know the complete factorization of $p - 1$:

$$p - 1 = q_1^{e_1} \ldots q_m^{e_m},$$

where $q_1 < \cdots < q_m$ are primes and $e_1, \ldots, e_m$ are *positive* integers. Note that if we cannot factor $p - 1$, then we cannot apply Gauss's method anyway. Since k divides $p - 1$, there exist non-negative integers $r_1, \ldots, r_m$ such that

$$k = q_1^{r_1} \ldots q_m^{r_m}$$

and $0 \le r_1 \le e_1, \ldots, 0 \le r_m \le e_m$.

Thus, in order to find k it is enough to compute $r_1, \ldots r_m$. Let's see how the process works for r_1; the other exponents are calculated by the same method. First, compute the sequence

$$a^{p-1}, a^{(p-1)/q_1}, a^{(p-1)/q_1^2}, \ldots, a^{(p-1)/q_1^{e_1}}$$

modulo p. Note that, by Fermat's theorem, the first element of the sequence is always 1. Suppose that w is the *biggest* non-negative integer such that

(6.1) $$a^{(p-1)/q_1^w} \equiv 1 \pmod{p}.$$

Then, either $w = e_1$ or

(6.2) $$a^{(p-1)/q_1^{w+1}} \not\equiv 1 \pmod{p}.$$

It follows from the key lemma and from (6.1) that k divides $(p - 1)/q_1^w$. On the other hand, we have from (6.2) that k does *not* divide $(p - 1)/q_1^{w+1}$.

In other words, $k = q_1^{r_1} \ldots q_m^{r_m}$ divides

$$q_1^{e_1 - w} q_2^{e_2} \ldots q_m^{e_m} = (p - 1)/q_1^w,$$

but does not divide

$$q_1^{e_1 - w - 1} q_2^{e_2} \ldots q_m^{e_m} = (p - 1)/q_1^{w+1}.$$

This can happen only if $r_1 = e_1 - w$, which gives us an algorithmic way to compute r_1.

Now let's return to the example. As mentioned in the proof of the primitive root theorem, it is convenient to choose $\overline{2}$ as the starting point for the application of Gauss's method. We must first compute the order of $\overline{2}$. In order to use the algorithm described above, we will factor the order of $U(41)$, which gives

$$\phi(41) = 40 = 2^3 \cdot 5.$$

Next we compute

$$\overline{2}^{40/2} = \overline{2}^{20} = \overline{1}$$

and also

$$\overline{2}^{40/2^2} = \overline{2}^{10} = \overline{40} \neq \overline{1}.$$

Thus the exponent of 2 in the order of $\overline{2}$ is $3 - 1 = 2$. Let's turn to the prime 5. A simple calculation shows that

$$\overline{2}^{40/5} = \overline{2}^8 = \overline{10} \neq \overline{1}.$$

Therefore the exponent of 5 in the order of $\overline{2}$ is $1 - 0 = 1$, and $\overline{2}$ has order $2^2 \cdot 5 = 20$. In particular, $\overline{2}$ is not a generator of $U(41)$.

Hence we must choose another element of $U(41)$, say $\overline{3}$. Once again it is necessary to compute its order, but

$$\overline{3}^{40/2} = \overline{40} \neq \overline{1} \quad \text{and} \quad \overline{3}^{40/5} = \overline{1}.$$

From the first equation we deduce that 8 divides the order of $\overline{3}$; from the second that 5 does not divide the order of $\overline{3}$. Hence, $\overline{3}$ has order 8.

Next we must factor $r = 20$ (the order of $\overline{2}$) and $s = 8$ (the order of $\overline{3}$) in the form of section 4. Since $r = 2^2 \cdot 5$ and $s = 2^3$, we can choose $r' = 5$ and $s' = 2^3$. This gives, in the notation of section 4, $u = 2^2$ and $v = 1$. Now, following the steps of the proof of the lemma in section 4, we construct the element

$$c = \overline{2}^u \cdot \overline{3}^v = \overline{2}^4 \cdot \overline{3}^1 = \overline{7}$$

of $U(41)$ that has order $r's' = 40$. Thus $\overline{7}$ is a generator of $U(41)$.

7. Exercises

1. Use the primality test of section 2 to show that 991 is a prime number.

2. Demonstrate that if $n > 0$ is an odd integer, and if 4 does not divide $n - 1$, then $(n - 1)^{(n-1)/2} \equiv 1 \pmod{n}$.

3. Use the test of section 2 to show that $M(7) = 2^7 - 1$ is prime.
Hint: If $p = 2^7 - 1$, then $2^7 \equiv 1 \pmod{p}$.

4. Use the test of section 2 and a computer algebra system to show that $R(23)$ is prime.

5. Let p be a prime number and let $n = 2p + 1$. Suppose that $2^{n-1} \equiv 1 \pmod{n}$ and that n is not divisible by 3.

    (1) Show that if q is a prime factor of n, then 4 has order p in $U(q)$.
    (2) Show that q is of the form $q = kp + 1$ for some integer $k > 0$.
    (3) Show that since $q < n$, then $k = 1$.
    (4) Combine these facts to show that n must be prime.
Hint for (1): $2^{n-1} = 4^p$. Hence, by hypothesis, $4^p \equiv 1 \pmod{n}$. If q is a prime factor of n, this congruence also holds for modulo q.

6. Show that:

    (1) If b is a prime number and $k \geq 3$ is an integer, then $b^{2^{k-2}} \equiv 1 \pmod{2^k}$.
    (2) $U(2^k)$ is *not* a cyclic group when $k \geq 3$.
Hint for (1): Use induction on k, beginning with $k = 3$.

7. The purpose of this exercise is to describe another primality test. It is based on *Wilson's theorem*, stated in (3) below.

    (1) Let G be a finite abelian group with an operation we will call multiplication. Show that the product of all the elements of G is equal to the product of those elements of G whose order is 2.
    (2) Let p be a prime. Show that the only element of order 2 of $U(p)$ is $\overline{-1} = \overline{p-1}$.

(3) Use (1) and (2) to show that $(p-1)! \equiv -1 \pmod{p}$. This result is known as *Wilson's theorem*.

(4) Show that if n is composite, then $(n-1)! \equiv 0 \pmod{n}$.

(5) Combining (3) and (4), we obtain the following primality test: A positive integer n is prime if and only if $(n-1)! \equiv -1 \pmod{n}$. Why isn't this an efficient way to test primality?

8. Use Gauss's method to find a generator for $U(73)$. This example is calculated by Gauss in section 74 of the *Disquisitiones arithmeticæ*.

9. Let $p > 0$ be an odd prime number.

(1) Show that if a is odd and $\bar{a}$ generates $U(p)$, then the class of a is a generator of $U(2p)$.

(2) Show that if a is even and $\bar{a}$ generates $U(p)$, then the class of $a+p$ is a generator of $U(2p)$.

(3) Show that $U(2p)$ is a cyclic group.

10. Let G be a finite cyclic group of order n generated by g. Let k be a positive integer.

(1) Show that g^k is a generator of G if and only if k is co-prime to n.

(2) Use (1) to show that G has $\phi(n)$ generators.

(3) How many generators does $U(p)$ have when p is prime?

11. Write a program to implement Pepin's test. The input will be the exponent $n \geq 0$; the output will be a message stating whether $F(n)$ is prime or composite. Note that the program will consist essentially of an implementation of the algorithm described in section 2 of the Appendix for computing powers in modular arithmetic. What is the biggest n for which your program can be used?

12. Write a program to implement the primality test of exercise 7. The program will consist essentially of an algorithm to find the residue of $(n-1)!$ modulo n. If you first compute $(n-1)!$ and then reduce it modulo n, the program will be applicable only to very small values of n. Instead, reduce modulo n at every step of the recursion that calculates $(n-1)!$. How long does it take for the program to stop when it is applied to the biggest prime smaller than 10^k for $k = 1, \ldots, 6$? Extrapolating from these results, try to determine how long your program would take to show that a number of 100 digits is prime.

.

11

The RSA cryptosystem

It is time to describe the RSA cryptosystem. Besides explaining how the system works, we must discuss its security in more detail; in other words, why is it so difficult to break a message encrypted using the RSA?

1. On first and last things

In order to implement the RSA cryptosystem for one user, it is necessary to choose two distinct prime numbers p and q, and compute $n = pq$. The primes p and q must be kept secret; the integer n will be a part of the *public key*. In section 5 we will discuss in detail a method for choosing these primes, and also how their choice is related to the security of the system.

Now, a message is encrypted by raising it to a power modulo n. So first we must find a way of representing the "plaintext" message as a set of classes modulo n. This is not really part of the encryption process; it is merely a way to prepare the message so that it can be encrypted.

To keep things as simple as possible, we will assume that the "plaintext" message contains only words written in uppercase letters. Thus the message is ultimately a sequence of letters and blank spaces. The first step consists of replacing each letter of the message by a number, using the following correspondence:

A	B	C	D	E	F	G	H	I	J	K	L	M
10	11	12	13	14	15	16	17	18	19	20	21	22

N	O	P	Q	R	S	T	U	V	W	X	Y	Z
23	24	25	26	27	28	29	30	31	32	33	34	35

The blank space between words is replaced by 99. Having done that, we obtain a number, possibly a very large one, if the message is long. However, it is not a number we want, but rather classes modulo n. Therefore, we must break the numerical representation of the message into a sequence of positive integers, each smaller than n. These are called the *blocks* of the message.

For example, the numerical representation of the motto "Know thyself" is

$$202,324,329,929,173,428,142,115.$$

If we choose the primes $p = 149$ and $q = 157$, then $n = 23{,}393$. Thus the numerical representation of the message above must be broken into blocks smaller than 23,393. One way to do this is as follows:

$$20{,}232 - 4329 - 9291 - 7342 - 8142 - 115.$$

Of course, the choice of blocks is not unique, but neither is it entirely arbitrary. For example, we cannot have blocks beginning with zero in order to avoid ambiguity at the decryption stage.

When an RSA-encrypted message is decrypted, one obtains a sequence of blocks. The blocks are then joined together to give the numerical representation of the message. It is only after replacing the numbers by letters, according to the table above, that one obtains the original message.

Note that we have made each letter correspond to a *two-digit number* in order to avoid ambiguities. For suppose that we had numbered the letters so that A corresponded to 1, B to 2, and so on. Then we wouldn't be able to tell whether 12 stood for AB or for the letter L, which is the twelfth letter of the alphabet. Of course, any convention that is unambiguous can be used instead of the one above. For example, one might prefer to use ASCII code, since the conversion of characters is automatically done by the computer.

2. Encryption and decryption

A message that has been prepared using the method of section 1 consists of a sequence of blocks, each one a number smaller than n. We must now explain how each block is encrypted. In order to do this we need n, the product of the two primes, and also another positive integer e, which must be invertible modulo $\phi(n)$. In other words, $\gcd(e, \phi(n)) = 1$. Note that it is easy to compute $\phi(n)$ if p and q are known; indeed,

$$\phi(n) = (p-1)(q-1).$$

The pair (n, e) is the *public* or *encryption* key of the RSA cryptosystem we are implementing. Let b be a block of the message; thus b is an integer and $0 \le b \le n - 1$. We will denote the block of the encrypted message that corresponds to b by $\mathbf{E}(b)$. The recipe for computing $\mathbf{E}(b)$ is the following:

$$\mathbf{E}(b) = \text{residue of } b^e \text{ modulo } n.$$

Note that each block of the message is encrypted separately. Thus the encrypted message is really a sequence of encrypted blocks. Moreover, we cannot reunite the blocks of the encrypted message into a number. If we do so, we will not be able to decrypt the message correctly. We will soon see why this is so.

Let's return to the example we considered in section 1. We chose $p = 149$ and $q = 157$, so that $n = 23{,}393$ and $\phi(n) = 23{,}088$. We must now choose e. Recall that e must be co-prime to $\phi(n)$. Since the smallest prime that does not divide 23,088 is 5, we can choose $e = 5$. Thus to encode the first block of the message of section 1 we must compute the residue of $20{,}232^5$ modulo 23,393. With the help of a computer algebra system we find that the residue is

20,036; hence $\mathbf{E}(20,232) = 20,036$. Encrypting the whole message, we have the following sequence of blocks:

$$20,036 - 23,083 - 11,646 - 4827 - 4446 - 13,152$$

Let's see how a block of the encrypted message is decrypted. In order to apply the decryption procedure we must know n and the inverse of e modulo $\phi(n)$. This last number is a positive integer we will denote by d. The pair (n, d) is called the *private* or *decryption* key of the RSA cryptosystem we are implementing. If a is a block of the encrypted message, then $\mathbf{D}(a)$ stands for the corresponding block of the decrypted message:

$$\mathbf{D}(a) = \text{residue of } a^d \text{ modulo } n.$$

Some comments are necessary before we return to the example. First, it is very easy to compute d when $\phi(n)$ and e are known. Indeed, it is a simple application of the extended Euclidean algorithm. Second, if b is a block of the original message, then we expect that $\mathbf{D}(\mathbf{E}(b)) = b$. In other words, decrypting a block of the encrypted message, we expect to find the corresponding block of the original message. Since this is not immediately obvious from the recipes given above, we give a detailed proof in the next section.

Finally, we have claimed in the introduction, and elsewhere in this book, that to break the RSA cryptosystem one needs to factor n, because it is necessary to know p and q in order to decrypt a message. Having described in detail how the system works, we have to face the fact that this claim is not quite correct. Besides n itself, we need only know d, the inverse of e modulo $\phi(n)$, to be able to apply the decryption procedure. Thus to break the system it is enough to compute d, when n and e are known. It turns out that this is equivalent to factoring n, as we will see in section 4.

In the example we have been discussing, $n = 23,393$ and $e = 5$. To compute d we apply the extended Euclidean algorithm to $\phi(n) = 23,088$ and 5.

remainders	quotients	x	y
23,088	*	1	0
5	*	0	1
3	4617	1	-4617
2	1	-1	4618
1	1	2	-9235

Thus $23,088 \cdot 2 + 5 \cdot (-9235) = 1$. Hence the inverse of 5 modulo 23,088 is -9235, and $d = 23,088 - 9235 = 13,853$, which is the smallest positive number congruent to -9235 modulo 23,088. Therefore, to decrypt the blocks of the encrypted message, we must raise them to the 13,853rd power modulo 23,393. In the example, the first encrypted block is 20,036. Calculating the residue of $20,036^{13,853}$ modulo 23,088, we conclude that $\mathbf{D}(20,036) = 20,232$. Note that even for such small numbers, the computations required to decrypt an RSA cryptogram are beyond the scope of most pocket electronic calculators.

3. Why does it work?

As we have already observed, the steps described above will constitute a practical cryptosystem only if, by applying the decryption procedure to a block of the encrypted message, we get the corresponding block of the original message. Suppose that we are considering an implementation of the RSA cryptosystem with encryption key (n, e) and decryption key (n, d). In the notation of section 2 we must show that if b is an integer and if $0 \leq b \leq n - 1$, then $\mathbf{DE}(b) = b$.

Actually, it is enough to prove that $\mathbf{DE}(b) \equiv b \pmod{n}$. To understand why, note that both $\mathbf{DE}(b)$ and b are non-negative integers smaller than n. Thus, if they are congruent modulo n, they must be equal. This explains why we need to break the numerical representation of the message into numbers smaller than n. It also explains why the blocks of the encoded message must be kept separate, otherwise the argument above will break down.

Now it follows from the recipes for encryption and decryption that

$$(3.1) \qquad \mathbf{DE}(b) \equiv (b^e)^d \equiv b^{ed} \pmod{n}.$$

However, d is the inverse of e modulo $\phi(n)$. Hence, there exists an integer k such that $ed = 1 + k\phi(n)$. Note that since e and d are integers greater than 2, and $\phi(n) > 0$, then $k > 0$. Replacing ed by $1 + k\phi(n)$ in (3.1), we obtain

$$b^{ed} \equiv b^{1+k\phi(n)} \equiv (b^{\phi(n)})^k b \pmod{n}.$$

Now Euler's theorem comes to our aid. Since $b^{\phi(n)} \equiv 1 \pmod{n}$, we have $b^{ed} \equiv b \pmod{n}$. Thus

$$\mathbf{DE}(b) \equiv b \pmod{n},$$

and the proof would be complete if it weren't for the fact that it is not quite correct.

If you reread the argument above with care, you will notice that we haven't taken into account the hypothesis of Euler's theorem. Indeed, in order to apply the theorem we must know that n and b are co-prime. This seems to imply that when breaking the message into blocks, we should make sure that the blocks are co-prime to n. Luckily this is not really necessary, because the congruence holds for any block. It is not the result we want to prove that is false, it is just that our proof of it is defective. The correct approach applies the argument used in the proof of Korselt's theorem in Chapter 6.

Recall that $n = pq$, where p and q are *distinct* positive primes. We will compute the residue of b^{ed} modulo p and modulo q. The computations are similar for both primes, so we will work out the details only for p. We have seen that

$$ed = 1 + k\phi(n) = 1 + k(p-1)(q-1)$$

for some integer $k > 0$. Therefore

$$b^{ed} \equiv b(b^{p-1})^{k(q-1)} \pmod{p}.$$

We want to apply Fermat's theorem, but we can only do this if p does not divide b. Suppose this is the case; then $b^{p-1} \equiv 1 \pmod{p}$, and we conclude that $b^{ed} \equiv b \pmod{p}$.

We have used Fermat's theorem instead of Euler's, but apparently we have the same problem as before: The congruence holds for some but not for all blocks. However, the blocks that have been left out are the ones divisible by p. Now if p divides b, then both b and b^{ed} are congruent to 0 modulo p. Thus the congruence holds in this case, too. Hence $b^{ed} \equiv b \pmod{p}$ holds for any integer b. Note that we could not have used a similar argument when we applied Euler's theorem to n. Indeed, $\gcd(n, b) \neq 1$ does not imply that $b \equiv 0 \pmod{n}$, because n is composite.

Thus we have proved that $b^{ed} \equiv b \pmod{p}$. A similar argument shows that $b^{ed} \equiv b \pmod{q}$. In other words, $b^{ed} - b$ is divisible by p and by q. But p and q are distinct primes, so $\gcd(p, q) = 1$. Thus, by the lemma of Chapter 2, section 6, pq divides $b^{ed} - b$. Since $n = pq$, we have $b^{ed} \equiv b \pmod{n}$ for any integer b. In other words, $\mathbf{DE}(b) \equiv b \pmod{n}$. As we pointed out at the beginning of the section, this is enough to prove that $\mathbf{DE}(b) = b$ because both sides of the equality are non-negative integers smaller than n. This shows that the recipes of the previous section give rise to a practical cryptosystem; we must now consider whether it is secure.

4. Why is it secure?

Recall that the RSA is a public key cryptosystem. The public key consists of $n = pq$, where p and q are distinct positive primes, and of another positive integer e, which is invertible modulo $\phi(n)$. Let's consider in detail what one has to do to break the RSA if all one knew was the pair (n, e).

In order to decrypt an RSA-encrypted block, we need to know $d > 0$, the inverse of e modulo $\phi(n)$. The problem is that, in practice, the only known way to do this is to apply the extended Euclidean algorithm to e and $\phi(n)$. However, to compute $\phi(n)$ by the formula of Chapter 8, section 4, we must know p and q, which confirms the original claim that to break the RSA we must factor n. Since this problem is, in general, very difficult, the RSA is secure.

However, we are free to imagine that someday, someone will invent an algorithm to compute d that does not require the knowledge of the factors of n. For example, what would happen if someone came up with an efficient algorithm to find $\phi(n)$ directly from n and e? This, it turns out, is just a disguised way of factoring n. In other words, if

$$n = pq \quad \text{and} \quad \phi(n) = (p-1)(q-1)$$

are known, then we can easily compute p and q. This is very easy to prove. Note first that

$$\phi(n) = (p-1)(q-1) = pq - (p+q) + 1 = n - (p+q) + 1,$$

so that $p + q = n - \phi(n) + 1$ is known. However,

$$(p+q)^2 - 4n = (p^2 + q^2 + 2pq) - 4pq = (p-q)^2,$$

so that $p - q = \sqrt{(p + q)^2 - 4n}$ is also known. But once we know $p + q$ and $p - q$, we can easily find p and q; thus we have factored n.

Therefore an algorithm to compute $\phi(n)$ is in fact an algorithm to factor n, and we are back at square one. However, that's not the end of it. We may go further and imagine that someone has invented an algorithm that finds d directly from n and e. But $ed \equiv 1 \pmod{\phi(n)}$. Thus, if we know n, e and d, then we know a multiple of $\phi(n)$. This is also enough to allow us to factor n. A probabilistic algorithm that does just this can be found in Koblitz 1987a, p. 91. In exercise 7 you will find a similar (but simpler) algorithm for factoring n, assuming that one can break *Rabin's cryptosystem*. This will give you an idea of what a probabilistic algorithm is like.

There is one last possibility: a method for finding the block b directly from the residue of b^e modulo n. If n is large enough, a systematic search for b among all possible candidates is out of the question, and no one has yet come up with any better idea. This is why it is widely believed that breaking the RSA cryptosystem is equivalent to factoring n, even though a proof of this fact is still lacking.

5. Choosing the primes

There is more to the security of the RSA than is apparent from the previous discussion. One important point has to do with the choice of the primes p and q. Of course, if they are small, the system is easy to break. But it is not even enough to choose large primes. Indeed, if p and q are large, but the difference $|p - q|$ is very small, then it is easy to factor $n = pq$ using Fermat's algorithm (see Chapter 2, section 4).

This is not idle talk. In 1995 two students of an American university broke a version of the RSA that was in public use. This was possible because of an unsuitable choice of primes for the system. On the other hand, the RSA has been in use for a long time and, if the primes are carefully chosen, it has proved to be very secure indeed. Thus an efficient method for choosing good primes is essential to the toolbox of anyone who intends to program the RSA.

Suppose we want to implement the RSA cryptosystem with a public key (n, e), such that n is an integer with approximately r digits. To construct n, choose the prime p with between $4r/10$ and $45r/100$ digits, and then choose q close to $10^r/p$. At present the recommended key size for personal use is 768 bits, which means that n will have approximately 231 digits. To construct such an n we will need two primes of, say, 104, and 127 digits. Note that these primes are far enough apart to make factoring n by Fermat's algorithm impracticable. However, we must also make sure that the numbers $p-1$, $q-1$, $p+1$, and $q+1$ do not have only small factors, because this would make n easy prey to some well-known factorization algorithms (see Riesel 1994, chapter 6, pp. 174–77). Let's now consider a method by which such large primes can be found.

First, however, we need a simple result on the distribution of primes. Recall that $\pi(x)$ stands for the number of positive primes less than or equal to x.

According to the prime number theorem, if x is large, then $\pi(x)$ is approximately equal to $x/\log x$, where $\log$ denotes the logarithm to base e; see Chapter 3, section 5. Now let x be a very large number and let ϵ be a positive number. We want an approximate value for the number of primes between x and $x + \epsilon$; that is, for $\pi(x + \epsilon) - \pi(x)$. It follows from the prime number theorem and the properties of logarithms that $\pi(x + \epsilon) - \pi(x)$ is approximately equal to

$$\frac{x + \epsilon}{\log x + \log(1 + x^{-1}\epsilon)} - \frac{x}{\log x}.$$

Assuming that $x^{-1}\epsilon$ is very small, we can replace $\log(1 + x^{-1}\epsilon)$ by zero and still get a reasonable approximation to $\pi(x + \epsilon) - \pi(x)$. We conclude that the number of primes between x and $x + \epsilon$ is approximately equal to $\epsilon/\log x$. Of course the *bigger* x is and the *smaller* ϵ is, the better the approximation. For a more detailed discussion see Hardy and Wright 1994, Chapter XXII, section 22.19.

Now suppose that we want to choose a prime near an integer x. For the sake of concreteness, suppose that x is of the order of magnitude 10^{127}. We will search for this prime in the interval x to $x + 10^4$. It would be helpful to know in advance how many primes are likely to be found in this interval. That's where the result of the previous paragraph comes to our aid. Note that, in this example, $x^{-1}\epsilon$ is of the order of magnitude 10^{-123}, which is indeed quite small. Thus, using the formula above, we conclude that in the interval x to $x + 10^4$ there are approximately

$$[10^4/\log(10^{127})] = 34$$

primes. At the end of Chapter 10, section 2, we sketched a strategy for proving that a given odd number n is prime. It consists of three steps:

(1) Check whether n is divisible by a prime smaller than 5000.
(2) Assuming that n is not divisible by any of these primes, apply Miller's test to n using as bases the first 10 primes.
(3) Assuming that the output of Miller's test to all these bases was "inconclusive", apply the primality test of Chapter 10, section 2, to n.

We will adapt this strategy to find a prime in the interval x to $x + 10^4$. First, we sieve the odd numbers in the given interval using the primes smaller than $5 \cdot 10^3$. Next, we apply (2) and (3) to each of the numbers left after sieving, until a prime is found.

To find out how much labor this entails, let's try to determine approximately how many integers will be left after the interval is sieved with the primes smaller than $5 \cdot 10^3$. Let m be a positive integer. If $x \leq km \leq x + 10^4$, then

$$[\frac{x}{m}] \leq k \leq [\frac{x + 10^4}{m}].$$

Thus there are

$$[\frac{x + 10^4}{m}] - [\frac{x}{m}]$$

multiples of m in the interval x to $x + 10^4$. This is approximately equal to $[10^4/m]$, which is also the number of positive multiples of m smaller than 10^4. This implies that the number of integers that are multiples of positive primes smaller than $5 \cdot 10^3$ in the intervals $[x, x + 10^4]$ and $[0, 10^4]$ is approximately the same. It is not difficult to compute the latter number. First, note that a composite number smaller than 10^4 must be a multiple of a prime smaller than $\sqrt{10^4} = 100$. Thus an integer in the interval $[0, 10^4]$ is a multiple of a prime smaller than $5 \cdot 10^3$ if it is composite or if it is itself a prime smaller than $5 \cdot 10^3$. Taking into account that 2 is the only even prime, we have that the number of *odd composite* integers smaller than 10^4 is $5000 - \pi(10^4) + 1$. Hence, the total number of odd integers left after sieving the interval x to $x + 10^4$ with the primes smaller than $5 \cdot 10^3$ is approximately equal to

$$5000 - (5000 - (\pi(10^4) - 1)) - (\pi(5 \cdot 10^3) - 1) = 560.$$

Note that $\pi(10^4)$ and $\pi(5 \cdot 10^3)$ are easily computed using the sieve of Erathostenes.

Thus we would expect to find, on average, 34 primes among a total of 560 integers left after the sieving process is applied to the given interval.

6. Signatures

If a company does its bank transactions by computer, it is clear that both the company and the bank will require that the information be encrypted before it is transferred between the computers. But this is not really enough. The bank must have some way of making sure that the message originated with a legitimate user at the company. The problem is that the bank's encryption key is public, so anyone can send the bank an encrypted message telling it, for instance, to transfer all the company's funds to the person's own account. How can the bank be sure that the message it has received is genuine? In other words, how can an electronic message be signed?

The method by which an electronic message is signed is quite simple and works for any public key cryptosystem. Let $\mathbf{E}_c$ and $\mathbf{D}_c$ be the company's encryption and decryption functions, and let $\mathbf{E}_b$ and $\mathbf{D}_b$ be the corresponding functions for the bank. Let a be a block of the message the company wishes to send the bank. We have seen that to make sure the message cannot be read by an eavesdropper, the company must send the bank the encrypted block $\mathbf{E}_b(a)$. To make sure that the message is also signed, the company will send the bank the block $\mathbf{E}_b(\mathbf{D}_c(a))$. In other words, the message is first encrypted using the company's private key, and the result is then encrypted again, this time with the bank's public key.

Having received the block $\mathbf{E}_b(\mathbf{D}_c(a))$, the bank will apply its decryption function to get $\mathbf{D}_c(a)$, and to this block it will then apply the company's encryption function to get a, the block of the original message. Note that $\mathbf{E}_c$ is public, and thus known to the bank.

Why is this enough to make sure that the message couldn't have originated outside the company? The bank must apply to the message it receives the sequence of functions

$$\mathbf{E}_c \mathbf{D}_b.$$

If the resulting message makes sense, the blocks of the original message must have been encrypted with the sequence

$$\mathbf{E}_b \mathbf{D}_c.$$

However, $\mathbf{D}_c$ is the decryption function of the company; thus it is secret, and its access is restricted to those employees who have the right of transacting business for the company. Needless to say, if a message is decrypted using $\mathbf{E}_c \mathbf{D}_b$ and makes sense, the probability that it has been encrypted using a function other than $\mathbf{E}_b \mathbf{D}_c$ is negligible.

There is one small drawback to applying the above method. Let (n_c, e_c) be the public key of the company, and (n_b, e_b) that of the bank. If a is a block of the original message, then $0 \le \mathbf{D}_c(a) < n_c$. Suppose now that $n_b < n_c$. In this case, we have no way to know in advance whether $\mathbf{D}_c(a)$ will be smaller than n_b. If it is bigger, it will be necessary to subdivide $\mathbf{D}_c(a)$ into blocks smaller than n_b before we apply $\mathbf{D}_b$ to it. If one does not do that, it is impossible to decrypt the resulting message correctly. This is called *reblocking*.

A simple way to avoid this problem is the following. Since both n_b and n_c are public, we can decide in advance which of them is smaller. If $n_c < n_b$, then we sign a message by encrypting its blocks using $\mathbf{E}_b \mathbf{D}_c$. However, if $n_b < n_c$, then we reverse the two functions and encrypt the blocks using $\mathbf{D}_c \mathbf{E}_b$. That way, the function corresponding to the smallest value of n always comes first, so that reblocking is unnecessary.

7. Exercises

1. Suppose that $n = 3,552,377$ is the product of two distinct primes, and that $\phi(n) = 3,548,580$; factor n.

2. The public key used by a bank in Toulouse to encode its messages using the RSA cryptosystem is $n = 10,403$ and $e = 8743$. Recently the computers at the bank received, from an unidentified source, the following message:

$$4746 - 8214 - 3913 - 9038 - 8293 - 8402$$

What does the message say?

3. The message

$$4199 - 215 - 355 - 1389$$

was encrypted using the RSA cryptosystem with public key $n = 7597$ and $e = 4947$. Moreover, it is known that $\phi(n) = 7420$. Decrypt the message.

4. Let p and q be odd primes, and suppose that we have an implementation of the RSA with public key (n, e), where $n = pq$. Now, a block b can be encrypted as itself under this implementation; in other words, it could happen that $\mathbf{E}(b) = b$. Such a block is said to be *fixed* under the RSA with key (n, e). Determine how many blocks are fixed under the RSA when $p = 3$, $q > 3$ and $e = 3$.

Hint: If b is fixed under the RSA with public key (n, e), then $b^e \equiv b \pmod{n}$. Thus $b^e \equiv b \pmod{p}$ and $b^e \equiv b \pmod{q}$. Solve each one of these equations when $p = 3$, $q > 3$, and $e = 3$, and use the Chinese remainder theorem.

5. Another well-known public key cryptosystem is the *El Gamal*. To construct this system it is necessary to choose a large prime p and a primitive root g modulo p. These will be common to all the users of a given implementation of the system. Now each user chooses a non-negative integer smaller than $p - 1$. This number must be kept secret, because it will be the decryption key of that user. The public key of a user, whose decryption key is a, will be $\overline{g}^a$. Let b be a block of the message to be encrypted; we must have $1 \le b \le p - 1$. To encrypt b, choose a random integer k and send the pair $(\overline{g}^k, \overline{b}\overline{g}^{ak})$.

    (1) Why is it easy to decrypt the message when both a and g are known?
    (2) What is required to break this cryptosystem?

The answer to (2) is known as the *discrete logarithm problem*. It is believed that this problem is as difficult to solve as that of factoring a large integer.

6. The cryptosystem invented by Michael Rabin in 1979 is very similar to the RSA. To begin, choose two distinct odd primes p and q, and let $n = pq$. Let b be a block of the message we want to encrypt; we must have $0 \le b < n$. The block b is encrypted as the residue of b^2 modulo n. If a is a block of the encrypted message, then we decrypt it by solving the equation $x^2 \equiv a \pmod{n}$. Let $u = (p^{q-1} - q^{p-1})$.

    (1) Show that $u^2 \equiv 1 \pmod{q}$ and $u^2 \equiv 1 \pmod{p}$.
    (2) Show that $u^2 \equiv 1 \pmod{n}$.
    (3) Show that, if x_0 is a solution of $x^2 \equiv a \pmod{n}$, then $-x_0$, ux_0 and $-ux_0$ are also solutions of the same equation.

Thus each block a of the cryptogram can be decrypted in four different ways, which is clearly a disadvantage.

7. Let $n = pq$, where p and q are distinct odd primes. We saw in section 4 that breaking the RSA is probably equivalent to factoring n. We will now prove that if one can find an algorithm that breaks Rabin's cryptosystem, then we have a probabilistic procedure for factoring n (see section 4). In other words, we want to show that if we have a machine that, having an integer a as input, outputs a solution of $x^2 \equiv a \pmod{n}$, then we can easily factor n and find p and q. First choose a random integer b and compute b^2 modulo n. Then use the machine to find a solution of $x^2 \equiv a \pmod{n}$. Since there are four solutions, there is one chance in two that the machine will find a solution x such that $x \not\equiv \pm b \pmod{n}$. Show that, in this case, $\gcd(x, b)$ must be either p or q. This means that, if we had a machine like this, we would expect to factor n after only two random choices of b.

8. Let p and q be primes that leave remainder 3 when divided by 4, and let $n = pq$. Suppose that a is a block of a message encrypted using Rabin's cryptosystem with public key n. Write a program that, having a and n as input, decrypts a using the method explained in exercise 6. In order to do this, the program will have to

    (1) factor n to find p and q, and
    (2) solve the equation $x^2 \equiv a \pmod{n}$.

Both stages have been the subject of previous exercises; namely, exercises 12 of Chapter 2, 19 of Chapter 5, and exercise 10 of Chapter 7. Use your program to decrypt the message below, which was encrypted using Rabin's cryptosystem with the public key $n = 20{,}490{,}901$:

$$2{,}220{,}223 - 18{,}957{,}657 - 11{,}291{,}133 - 2{,}180{,}507 - 41{,}1224{,}784$$

You'll obtain $4^5 = 1024$ possible decryptions, but only one of them will make sense.

Coda

In his *Principles of Human Knowledge*, published in 1710, George Berkeley had the following to say about number theory and its practitioners:

> The opinion of the pure and intellectual nature of numbers in abstract, hath made them in esteem with those philosophers, who seem to have affected an uncommon fineness and elevation of thought. It hath set a price on the most trifling numerical speculations which in practice are of no use, but serve only for amusement ... .

A few lines later, he adds

> we may perhaps entertain a low opinion of those high flights and abstractions, and look on all enquiries about numbers, only as so many *difficiles nugae*, so far as they are not subservient to practice, and promote the benefit of life.

By the way, the Latin expression *difficiles nugae* means "difficult trivialities".

Berkeley's opinion that number theory was the most useless of all types of mathematics survived into the twentieth century. Thus G. H. Hardy, after noting that science may be used for good or evil ends, says,

> both Gauss and lesser mathematicians may be justified in rejoicing that there is one science at any rate, and that their own, whose very remoteness from ordinary human activities should keep it gentle and clean.

The quotation is from Hardy's famous *A Mathematician's Apology* (Hardy 1988, p. 120). But we ought not to forget that it was written during the Second World War by one who always remained a confirmed pacifist. How Hardy would have reacted to the applications of number theory described in this book we can only guess. But one thing is certain: Modern cryptography completely changed our views toward the applicability of number theory.

There is no better way to finish this book than by pointing to topics of both number theory and cryptography that have not been discussed in this book, and that may serve as directions for further study.

An obvious absence in the book is a serious discussion of the cryptoanalysis of the RSA. In other words, of the ways by which one can go about breaking a given implementation of the RSA cryptosystem. Most of the research in this area is related to methods for factoring the public key. Moreover, in recent years, some spectacular factorization algorithms have been discovered, among them *Lenstra's elliptic curve algorithm*, the *quadratic sieve*, and the *number field sieve*.

An *elliptic curve* is a nonsingular plane cubic curve; that is, a curve described by a polynomial equation of degree 3 in two variables that has a well-defined tangent at each of its points. For example, the points (x, y) of the plane that satisfy the equation $y^2 = x^3 + 17$ form an elliptic curve. As a consequence of the fundamental theorem of algebra, most lines drawn through two points of an elliptic curve will intersect the curve into a third point. Using this fact, one can define an operation, called addition, on the points of the curve. Furthermore, the points of the curve form a group under this addition.

Let's assume from now on that the elliptic curves we are considering are defined by polynomials with integer coefficients. A *rational point* of such a curve is a point of the plane that belongs to the curve and whose coordinates are rational numbers. Thus $(-1, 4)$ is a rational point of the elliptic curve whose equation is $y^2 = x^3 + 17$. One can prove that the set of rational points of an elliptic curve is a subgroup of the group of points of the whole curve. Thus, given two rational points of an elliptic curve, we can obtain another one by adding them up.

The problem of finding rational points on elliptic curves is closely related to the diophantine problems discussed in this book. This problem has been studied by many famous mathematicians, and it has been at the forefront of developments in number theory. Elliptic curves are also central to A. Wiles's solution of Fermat's Last Theorem. In fact, what Wiles proved was a famous conjecture about elliptic curves that had been shown to imply Fermat's statement (see Gouvêa 1994).

In a paper published in 1987, H. W. Lenstra showed that the group properties of an elliptic curve can be used to factor large numbers. Lenstra's algorithm is most effective for integers that are difficult to factor by trial division, but that have less than 30 digits. Luckily a masterful, but truly elementary, introduction to elliptic curves is now available that includes a detailed discussion of Lenstra's algorithm. It is Silverman and Tate's *Rational Points on Elliptic Curves* (1992). Lenstra's algorithm appeared originally in Lenstra 1987; it is also discussed in Bressoud 1989 and in Cassels 1991.

The second factorization algorithm mentioned above, the quadratic sieve, is best used for numbers too large to be tackled by Lenstra's algorithm. The quadratic sieve was the algorithm used to factor the public key of the RSA-129 challenge mentioned in section 2 of the Introduction. This algorithm has the great advantage of making it very easy to distribute the task of factoring to many computers. Thus people can volunteer the idle time of their personal computers for use in these mammoth factorization efforts. Of course, none of this would be possible, at least on the scale on which it was done for the RSA-129, before the advent of the internet.

The quadratic sieve grew out of an enhancement of Fermat's algorithm proposed by M. Kraitchik in the 1920s. It makes use of the ideas and methods related to a theorem that is surely a great favorite of all number theory lovers, the *law of quadratic reciprocity*, proved by Gauss in 1796. Many mathematicians suggested improvements to Kraitchick's method, and these led to C. Pomerance's proposal of the quadratic sieve in 1981. The word *sieve* is not used here for nothing. Indeed, Pomerance's key idea to speed up the method was to use a sieve very similar to Erathostenes'.

An elementary approach to the quadratic sieve can be found in Bressoud 1989. It includes all the necessary prerequisites, and a proof of the law of quadratic reciprocity to boot. The story of the sieve is told by Pomerance himself in Pomerance 1996.

The number field sieve brings us into the world of quadratic number fields, a less elementary topic than the ones mentioned hitherto. The original idea was circulated in a letter of J. Pollard in 1988. The first big prize collected by this algorithm was the factorization of the ninth Fermat number, a 155-digit number, in 1990; see Lenstra et al. 1993. This paper contains a description of the number field sieve, which is also discussed in Pomerance 1996. The factorization of the RSA-130 in April 1996 was done with a further improvement to this algorithm. For details on quadratic number fields see Ireland and Rosen 1990, Chapter 13.

Before we move on to topics beyond the RSA, let's make it clear that, in practice, breaking this cryptosystem is not the same as factoring the public key. The best evidence of this fact to date was provided in 1995 by P. Kocher, an independent security consultant who had just obtained an undergraduate degree in biology. He showed that it is possible to break some versions of the RSA by using information about the length of time it takes a legitimate recipient of the message to decrypt it; see Kocher 1996. This shows that the security of the RSA, and of other cryptosystems, is not merely a question of developing better algorithms and more powerful computers.

As we saw in Chapter 11, the RSA is not the only cryptosystem suggested by number theoretic problems that are hard to solve. One of the examples mentioned, the El Gamal system, has also been used. Another system of the same vintage was invented by N. Koblitz in 1987. In it, the computation of multiples of a point in an elliptic curve replaces the exponentiation modulo n that occurs in both RSA and El Gamal. For Koblitz's original proposal see Koblitz 1987b.

Finally, number theory still thrives on the beautiful and mostly useless results so harshly condemned by George Berkeley. The most treasured elementary book on number theory is surely Hardy and Wright's classic *Introduction to the Theory of Numbers*. In it you'll find a proof of Gauss's law of quadratic reciprocity, a study of partitions of integers, a proof of Fermat's Last Theorem for $n = 3$, a lot of information on quadratic fields, and many elementary facts about the distribution of primes.

Many people have learned number theory from the above book. The mathematical physicist Freeman Dyson, who won the book as a prize at age 14, was one. Here is what he had to say about Hardy and Wright in a 1994 interview (see Albers 1994, p. 7):

> It's a marvelous book. It's the finest book I know on number theory. It's not a textbook, but it's an enormously readable account of the subject, written with such style that it's probably the best introduction to mathematics that has ever been written.

A more modern introduction to number theory is Ireland and Rosen's *Classical Introduction to Modern Number Theory*. It contains very recent developments, but you'll need to have a good grasp of modern algebra (groups, fields, and rings) to read even the more elementary chapters.

Finally, for those with a very strong historical bent, there is A. Weil's *Number Theory: An Approach through History*. Written by an acknowledged master of number theory, who is also well known for his work on the history of mathematics, it deals with the contributions of Fermat, Euler, Lagrange, and Legendre. This is a fascinating book, and one you'll probably often go back to for the sheer delight in the feast it offers.

Appendix

Roots and powers

In this Appendix we describe two algorithms required for the implementation of the factorization algorithms and primality tests presented in this book. The algorithm of section 1 computes the integer part of the square root of a given positive integer; the algorithm of section 2 computes powers in modular arithmetic.

1. Square roots

Both of the factorization algorithms in Chapter 2 require the computation of square roots. In this section we describe a procedure that can be used to compute the integer part of the square root of a positive integer. This is exactly what is required for the trial division algorithm of Chapter 2, section 2. However, in the case of Fermat's algorithm, what is needed is a procedure to determine whether a given positive integer n is a perfect square. But n is a perfect square if and only if $n - [\sqrt{n}]^2 = 0$. So the algorithm of this section can also be used to settle this question.

The procedure consists of computing a decreasing sequence of positive integers

$$x_0, x_1, x_2, \ldots$$

such that, if $x_k \leq x_{k+1}$ for some $k \geq 0$, then $x_k = [\sqrt{n}]$. The sequence is defined recursively by

$$x_0 = \left[\frac{n+1}{2}\right] \quad \text{and} \quad x_{i+1} = \left[\frac{x_i^2 + n}{2x_i}\right]$$

for $i \geq 0$. The idea is that, starting at $(n+1)/2$, the sequence will decrease until it reaches $[\sqrt{n}]$. Now consider the following statements:

(1) $[(n+1)/2] \geq [\sqrt{n}]$;
(2) if $x_k > [\sqrt{n}]$ then $x_k > x_{k+1}$; and
(3) if $x_k \leq x_{k+1}$ then $x_k = [\sqrt{n}]$.

Thus each x_k is a positive integer bounded above by $(n+1)/2$. Hence there can be only finitely many distinct elements in the sequence. In particular, there must be some r for which $x_r \leq x_{r+1}$, and (3) then implies that $x_r = [\sqrt{n}]$.

Therefore, we need only prove the three statements above. Note that if $y \geq x$ are real numbers, then $[y] \geq [x]$. Thus (1) follows if we prove that

$(n+1)/2 > \sqrt{n}$. But this last inequality is an immediate consequence of

$$\frac{(n+1)^2}{2} = \frac{n^2+2n+1}{2} > n^2.$$

To prove (2) recall that, by definition, $x_{k+1} = [(x_k^2+n)/2x_k]$. Since x_k is an integer, it follows from $x_k > [\sqrt{n}]$ that $x_k > \sqrt{n}$. Squaring both sides of the inequality, we have $x_k^2 > n$. Hence $2x_k^2 > x_k^2 + n$, so that

$$x_k > \frac{x_k^2+n}{2x_k} \geq x_{k+1}.$$

To prove (3), suppose that $x_k \leq [(x_k^2+n)/2x_k]$. This is equivalent to

$$0 \leq \frac{x_k^2+n}{2x_k} - x_k < 1,$$

which implies

$$0 \leq n - x_k^2 < 2x_k.$$

Hence $n \geq x_k^2$ and $n < x_k^2 + 2x_k$. But from this last inequality it follows that $n < (x_k+1)^2$. Therefore

$$x_k^2 \leq n < (x_k+1)^2,$$

which implies that $x_k = [\sqrt{n}]$, thus proving (3).

This method of finding the integer part of the square root is easily implemented in the following algorithm.

Square root algorithm

Input: an integer $n > 2$
Output: the integer part of the square root of n
Step 1 Begin by setting $X = n$ and $Y = [(n+1)/2]$, and go to step 2.
Step 2 If $X \leq Y$, stop and write X; otherwise go to step 3.
Step 3 Replace the value of X by that of Y, and the value of Y by $[(X^2 + n)/2X]$, and return to step 2.

2. Power algorithm

Suppose that we have three positive integers a, e, and n. In this section we describe an algorithm to find the residue of a^e modulo n. This algorithm uses the binary expansion of the exponent to compute the power in a very efficient way. It is also very easy to implement.

Suppose that

$$e = b_n 2^n + \cdots + b_1 2 + b_0,$$

where the coefficients $b_0, b_1, \ldots, b_n$ are either 0 or 1. Thus we have

$$a^e = (a^2)^{b_n 2^{n-1} + \cdots + b_2 2 + b_1} \cdot a^{b_0}.$$

Note that a^{b_0} can be 1 (if $b_0 = 0$) or a (if $b_0 = 1$). If $P_1 = a^{b_0}$, then

$$a^e = (a^4)^{b_n 2^{n-2} + \cdots + b_3 2 + b_2} \cdot (a^2)^{b_1} P_1.$$

Now let $P_2 = (a^2)^{b_1} P_1$; thus

$$a^e = (a^8)^{b_n 2^{n-3} + \cdots + b_4 2 + b_3} \cdot (a^4)^{b_2} P_2.$$

Continuing this, we obtain a sequence of integers $P_1, P_2, \ldots P_n$, where $P_n = a^e$. Of course, if we are computing in $\mathbb{Z}_n$, we will reduce each product modulo n at every step of the calculation.

Note that, at every step, either we square a number or we compute the product $a^{2^i} P_i$ for $i = 1, 2, \ldots, n$. Moreover, if at step i we have $b_i = 0$, then it is not necessary to compute $a^{2^i} P_i$.

In practice the algorithm finds the binary expansion of e while it is computing the power. Thus, if e is odd, then $b_0 = 1$; if e is even, then $b_0 = 0$. We can find b_1 by applying a similar procedure to

$$b_n 2^{n-1} + \cdots + b_2 2 + b_1.$$

Note that this last number is equal to $e/2$, if e is even, and to $(e-1)/2$, if e is odd. And so on. The algorithm is the following.

Power algorithm

Input: integers a, e, and n, where $a, n > 0$ and $e \geq 0$
Output: the residue of a^e modulo n
Step 1 Begin by setting $A = a$, $P = 1$, and $E = e$.
Step 2 If $E = 0$, write "$a^e \equiv P \pmod{n}$"; otherwise go to step 3.
Step 3 if E is odd, then give P the value of the residue of $A \cdot P$ modulo n and E the value $(E-1)/2$, and go to step 5; otherwise go to step 4.
Step 4 If E is even, then give E the value $E/2$ and go to step 5.
Step 5 Replace the present value of A by the residue of A^2 modulo n and go to step 2.

Bibliography

Adleman, L., Rivest, R. L., and Shamir, A.

> **1978** A method for obtaining digital signatures and public-key cryptosystems. *Comm. ACM* **21**, 120–126.

Akritas, A. G.

> **1989** *Elements of computer algebra with applications.* John Wiley & Sons, New York.

Albers, D. J.

> **1994** Freeman Dyson: Mathematician, physicist, and writer. *The College Mathematics J.* **25**, 3–21.

Alford, W. R., Granville, A., and Pomerance, C.

> **1994** There are infinitely many Carmichael numbers. *Ann. of Math.* **140**, 703–722.

Arnault, F.

> **1995** Constructing Carmichael numbers which are strong pseudoprimes to several bases. *J. Symbolic Computation* **20**, 151–161.

Artin, M.

> **1991** *Algebra.* Prentice-Hall, Englewood Cliffs.

Bateman, P. T., and Diamon, H. G.

> **1996** A hundred years of prime numbers. *The Amer. Math. Monthly* **103**, 729–741.

Bressoud, D. M.

> **1989** *Factorization and primality testing.* Undergraduate Texts in Mathematics. Springer-Verlag, New York.

Bruce, J. W.

> **1993** A really trivial proof of the Lucas-Lehmer test. *The Amer. Math. Monthly* **100**, 370–371.

Carmichael, R. D.

1912 On composite numbers P which satisfy the Fermat congruence $a^{P-1} \equiv 1$ (mod P). *The Amer. Math. Monthly* **19**, 22–27.

Cassels, J. W. S.

1991 *Lectures on elliptic curves.* London Math. Soc. Student Texts 24. Cambridge University Press, Cambridge.

Davies, W. V.

1987 *Egyptian hierogliphs.* British Museum Publications, London.

Davis, M.

1980 What is computation? In *Mathematics Today*, edited by L. A. Steen. Vintage Books, New York, 241–267.

Dickson, L. E.

1952 *A history of the theory of numbers.* Chelsea Publishing Company, New York.

Edwards, H. M.

1977 *Fermat's last theorem.* Graduate Texts in Mathematics 50. Springer-Verlag, New York.
1984 *Galois theory.* Graduate Texts in Mathematics 101. Springer-Verlag, New York.

Gauss, C. F.

1986 *Disquitiones Arithmeticæ.* Translated by A. A. Clarke. Revised by W. C. Waterhouse with the help of C. Greither and A. W. Grotendorst. Springer-Verlag, New York.

Giblin, P.

1993 *Primes and programming.* Cambridge University Press, Cambridge, England.

Gostin, G. B.

1995 New factors of Fermat numbers. *Math. of Comp.* **64**, 393–395.

Gouvêa, F. Q.

1994 A marvellous proof. *The Amer. Math. Monthly* **101**, 203–222.

Hardy, G. H.

1963 *A course of pure mathematics.* 10th edition. Cambridge University Press, Cambridge, England.

Hardy, G. H.

   1988 *A mathematician's apology.* Cambridge University Press, Cambridge, England.

Hardy, G. H., and Wright, E. M.

   1994 *An introduction to the theory of numbers.* 5th edition. Oxford Science Publications. Oxford University Press, Oxford, England.

Ingham, A. E.

   1932 *The distribution of primes.* Cambridge University Press, Cambridge, England.

Ireland, K., and Rosen, M.

   1990 *A classical introduction to modern number theory.* 2nd edition. Graduate Texts in Mathematics 84. Springer-Verlag, New York.

Jaeschke, G.

   1993 On strong pseudoprimes to several bases. *Math. of Comp.* **61**, 915–926.

Kang Sheng, S.

   1988 Historical development of the Chinese remainder theorem. *Arch. Hist. Exact Sci.* **38**, 285–305.

Knuth, D. E.

   1981 *The art of computer programming.* Vol. 2, *Seminumerical algorithms.* 2nd edition. Addison-Wesley Publishing Company, Reading, Massachusetts.

Koblitz, N.

   1987a *A course in number theory and cryptography.* Graduate Texts in Mathematics 97. Springer-Verlag, New York.
   1987b Elliptic curve criptosystems. *Math. of Comp.* **48**, 203–209.

Kocher, P.

   1996 Timing attacks on implementations of Diffie-Helman, RSA, DSS, and other systems. In *Advances in Cryptology-CRYPTO '96*, edited by N. Koblitz. Lecture Notes in Computer Science 1109. Springer-Verlag, 104–113.

Kranakis, E.

   1986 *Primality and criptography.* Wiley-Teubner Series in Computer Science. B. G. Teubner and J. Wiley & Sons.

Lenstra, A. K., Lenstra, H. W., Jr., Manasse, M. S., and Pollard, J. M.

   1993 The factorization of the ninth Fermat number. *Math. of Comp.* **61**, 319–349.

Lenstra, H. W.

   1987 Factoring integers with elliptic curves. *Math. of Comp.* **126**, 649–673.

Plato

    1982 *Plato's Republic.* Translated by B. Jowet. Modern Library, New York.

Poincaré, H.

    1952 *Science and hypothesis.* Dover, New York.

Pomerance, C.

    1996 A tale of two sieves. *Notices of the Amer. Math. Soc.* **43**(12), 1473–1485.

Pomerance, C., Selfridge, J. L., and Wagstaff, S. S., Jr.,

    1980 The pseudoprimes to $25 \cdot 10^9$. *Math. of Comp.* **151**, 1003–1026.

Rabin, M. O.

    1980 Probabilistic algorithm for testing primality. *J. Number Theory* **12**, 128–138.

Ramanujan, S.

    1927 *Collected papers of Srinivasa Ramanujan.* Edited by G. H. Hardy, P. V. Seshu Aiyar, and B. M. Wilson. Cambridge University Press, Cambridge, England.

Ribenboim, P.

    1990 *The book of prime number records.* Springer-Verlag, New York.
    1994 *Catalan's conjecture.* Academic Press, Boston.

Riesel, H.

    1994 *Prime numbers and computer methods of factorization.* 2nd edition. Progress in Mathematics 126. Birkhäuser, Boston.

Rigatelli, L. T.

    1996 *Evariste Galois.* Translated by J. Denton. Vita Mathematica. Birkhäuser, Basel.

Rotman, J. J.

    1984 *An introduction to the theory of groups.* 3rd edition. Allyn and Bacon, Newton, Massachusetts.

Silverman, J. H., and Tate, J.

    1992 *Rational points on elliptic curves.* Undergraduate Texts in Mathematics. Springer-Verlag, New York.

Thomas, I., trans.

    1991 *Greek Mathematical Works I: Thales to Euclid.* Loeb Classical Library 335. Harvard University Press, Cambridge, Massachusetts, and London, England.

van der Waerden, B. L.

    1985 *A history of algebra.* Springer-Verlag, Berlin–Heidelberg.

Weil, A.

    1987 *Number theory: An approach through history.* Birkhäuser, Boston.

Weyl, H.

    1982 *Symmetry.* Princeton University Press, Princeton.

Index of the main algorithms

Algorithm	Aim	Page
Division algorithm	Computes quotient and remainder	20
Euclidean algorithm	Computes greatest common divisor	23
Extended Euclidean algorithm	Computes greatest common divisor and coefficients of the linear combination	27, 30
Factorization by trial division	computes smallest factor of a given positive integer	35
Fermat's algorithm	Computes a factor of a given integer	38
Sieve of Erathostenes	Finds all primes smaller than a given bound	60
Miller's test	Determines whether a number is composite	101
Chinese remainder algorithm	Solves linear systems of congruences	110, 113
Fermat's method	Finds a factor of a Mersenne number with prime exponent	142
Euler's method	Finds a factor of a Fermat number	144
Lucas-Lehmer test	Determines whether a given Mersenne number is prime or composite	147
Lucas's test	Determines whether a given number is prime	151
Pepin's test	Determines whether a given Fermat number is prime or composite	152
Primality test	Determines whether a given number is prime	154
Gauss's method	Finds a primitive root modulo a prime p	158

Index of the main results

Algorithm	Content	Page
Division theorem	Existence and uniqueness of quotient and remainder	21
Unique factorization theorem	Every integer can be uniquely written as a product of prime powers	33
Fundamental property of primes	If a prime divides a product, then it divides one of the factors	41
Invertibility theorem	Existence of inverses modulo n	75
Principle of finite induction	Method of proof	84, 90
Fermat's theorem	If p is prime, then $a^p \equiv a \pmod{p}$	86, 88
Korselt's theorem	Characterization of Carmichael numbers	99
Chinese remainder theorem	Solution of linear systems of congruences	111, 114
Lagrange's theorem	The order of a subgroup divides the order of the group	132, 136
Euler's theorem	$a^{\phi(n)} \equiv 1 \pmod{n}$ if $\gcd(a, n) = 1$	135
Key lemma	$a^k = e$ in a group if and only if the order of a divides k	141
Primitive root theorem	If p is prime, then the group $U(p)$ is cyclic	151/158

Index